THE WASHINGTON EARTHQUAKE HANDBOOK:

AN EASY-TO-UNDERSTAND INFORMATION AND SURVIVAL MANUAL

BY

VERN COPE

PUBLISHED BY VERN COPE
PORTLAND, OREGON

This book is dedicated to
Linda Lawrance Noson and Brian Atwater,
dedicated pioneers

Also by Vern Cope

THE OREGON EARTHQUAKE HANDBOOK

Library of Congress Catalog Card Number 93-91101
ISBN 0-9635564-4-4

Book designed and produced by Mimi Yahn
Manufactured in the United States of America
First Edition, 1994

Published by Vern Cope
P.O. Box 19843
Portland, OR 97280

Printed on recycled paper using soy-based ink

TABLE OF CONTENTS

LIST OF ILLUSTRATIONS

ACKNOWLEDGEMENTS

This book is the first publication in some years which combines up-to-date information for the general public about Washington's geology, how individuals can get ready for earthquakes and information about what government and private groups are doing in preparation. Because the states of the Pacific Northwest have similar earthquake faults and fault zones and because these are significantly different from those in California, I decided several years ago to publish handbooks for both Oregon and Washington, each with material relevant to that state. This book is a companion volume to The Oregon Earthquake Handbook, published in 1993.

The information summarized here has been designed for individuals and families, but not for businesses. It was obtained from extensive research which included books, pamphlets, video tapes, scientific and popular magazines, and from taped interviews.

Library research was conducted primarily at Portland State University, the Multnomah Country Central Library (Portland) and at the well-stocked Division of Geology and Earth Resources Library (Washington State Department of Natural Resources) in Olympia, where Connie Manson and her staff demonstrated impressive knowledge, organization and efficiency. This researcher cannot thank them adequately.

I wish to thank the following Washingtonians and Oregonians for allowing me to tape interviews with them: Scott Carpenter, Washington Insurance Council, Bellevue; Mark Darienzo, Geology Department, Portland State

Unversity; Teri Giles, Disaster Services, Seattle-King County Chapter, American Red Cross, Seattle; Lloyd Marbet, Oregon Conservancy Foundation; Darlena Wilson, Grays Harbor Department of Emergency Services, Montesano; and the following individuals with the Washington State Department of Community Development, Olympia: Max Messman, Susan Larson and John Vollmer, Emergency Management Division, and Willy O'Neil, Washington State Building Code Council.

Several individuals deserve special recognition:

First to Matthew Mabey of the Oregon Department of Geology and Mineral Industries and James Bela, Oregon Earthquake Awareness, who went well beyond the requirements of an interview by extensively critiquing early drafts of this work and taking the time to explain some very important concepts which had eluded me in my research.

Second, and most importantly, to one of the loves of my life, my wife Mimi Yahn, whose patient and skillful editing made this a much more understandable and readable book. Mimi also drew all the illustrations and is responsible for the design and desktop publishing of this work.

In addition, thanks to: two other loves of my life, my daughter, Lily Copenagle, and my mother, Hazel Cope, for their helpful feedback; Steve Corbin, Building Inspection Service, Inc., Metairie, Louisiana; and to Robert Ward of Bellevue, Peter May, University of Washington, Seattle, and to Mothers For H.E.L.P., Kirkland, among others.

It should be noted that this is my work and that any errors within are mine and mine alone.

A FINAL NOTE: The author is neither a geologist nor an expert in emergency management. Although I have worked hard to accurately describe the best actions to be taken before, during and after earthquakes, I take no responsibility or liability for the actions taken by anyone who uses the material in this manual.

INTRODUCTION:
DEATH, TAXES AND THE BIG ONE
OR
LIFE ON THE EDGE

We in the Pacific Northwest rarely feel threatened by our natural environment. Our weather is generally temperate and seldom are we subject to the extremes of tropical heat or arctic cold. Few of us are affected by flooding, tornados or hurricanes and we gloat that the possibility of major earthquakes will only keep Californians awake at night. Aren't we glad we don't live in California? Well, unfortunately, we don't! The risk we face from earthquakes here is actually *greater than* faced by many of our neighbors to the south!

Although there has been fairly extensive coverage of the potential for great Northwest quakes in some newspapers and on local T.V. news programs, only a small percentage of us seem to be aware of the danger, and an even smaller percentage take this danger seriously. Ignorance is bliss. And ignorance is dangerous. A major purpose of this book is to convince you of the very real possibility of devastating earthquakes in Washington and to point out some problems with preparedness for these events which may well happen in our lifetimes. In addition, it is the purpose of this book to show how we can, as individuals, as families and as communities, prepare ourselves to deal with the devastation that a major quake will undoubtedly create.

We must be prepared physically and psychologically. To not deal intelligently with what is now known about our precarious situation would be the height of folly.

It's also very important to get the word out to a very wide audience. *Everyone* in western Washington will be affected by a great quake. For this reason, all discussion will be in easy-to-understand terms, so that each of us can easily convey the information to others without numbing them with scientific jargon. I hope this book will be an aid to those who realize Washington needs a greater statewide citizen and governmental awareness to deal with this impending crisis and to those who decide to become activists in this crucial area.

Finally, you'll find some humor in this book. Most humor, of course, is based on misfortune and tragedy. But without humor, we may take this information too seriously and run the risk of denial, inaction and despair.

PART ONE

THE BIGGEST THING TO HIT WASHINGTON IN A LONG TIME:

THE PROBLEM

CHAPTER 1

TRUE STORIES

A STORY OF THE PAST

She had spent much of the morning at her family's area in the largest of the cedar longhouses, working a loom to make a blanket for her children. As she was carefully admiring her creation made from hair, cedar bark and cattail, her son, who had been playing at her side with his small canoe and harpoon toys, startled her. "Listen," he said, "they're back and they have seal!"

He ran out of the longhouse and she followed, walking briskly with the excitement all the village was feeling at that moment. As she pushed aside the mat door, she saw all the people standing next to a canoe. The length of five men, it held the carcasses of two beautiful fur seals. She ran back into the longhouse and in a minute had changed from her cedar bark skirt into one made of cattail reeds for her new labor.

Soon she was sweating with the other women, preparing the seals for smoking. Far down the beach she saw her son, still clutching his toys, with her husband, probably talking of the hunt.

Now, as she works on the oily meat, she notices something out of the corner of her eye at the same moment as she has a strange feeling in her body—she sees a tall tree on the cliff above the village sway with no wind and her legs are shaken for a second by something under her bare feet.

Filled with the excitement of the moment, no one else

seems to notice. She looks around and laughs nervously for a few seconds but stops in terror when she feels the sand beneath her shake wildly. She looks around for something to hold onto but everything is moving and falling. The sand moves like the waves in the sea and it seems to move without an end. She lies for a long time on the beach crying, then screaming for it to stop.

For a few moments, she is able to see the village. Many of the houses have collapsed, the canoes at the water's edge have overturned and the sandy shore has become liquid. She sees people helplessly trying to move in her direction, trying to move away from the cliff which is cascading straight down on them. She desperately tries to stand up to get to her son but she faints.

When she regains consciousness, she sees an even more amazing and more horrifying thing happening in the ocean. An immense wave, the full length of the coast, rises high, much higher than the largest wave of any storm, and rushes toward her. It pounds against her, then sweeps her against a large rock, smashing her dead.

As the huge wave of water and bodies and wood plows into the cliff, more of the soil collapses, burying what is left of the village.

A few dazed survivors cling to overturned canoes. Soon they will begin the long process of rebuilding the village and preparing for winter. In time, as it has been for thousands of years, the only memory of the great waves and shaking ground will be in the stories of the generations to follow.

A STORY OF THE FUTURE

It's getting late. You have to leave soon, but there's still time to finish your coffee with no rush. Your spouse and children left ten minutes ago. Since it's not your day to take the kids to school, you have some extra time to read the paper. But why bother—it's all bad news: it seems that another one of the confederation's republics never did turn over all of its nuclear weapons to Russia, the Tokyo Stock Market is in free fall and more problems are expected in your industry which might mean you'll be the next to be "surplused." And that's just the front page! Doom and gloom. Better to not read the paper anymore. Ignorance is bliss.

Then, for no apparent reason, the chairs and the table bounce for just a second. You pause, then smile, not sure if this really happened. Uneasily, you go back to reading.

Oops. You notice your watch shows 7:57. Time to get going. You bring the coffee cup to your lips for a last swig and begin to stand up at the same moment. You suddenly feel unsteady, your legs buckle slightly. You hear what seems to be a truck driving by. But there's no truck outside that you can see.

Then you catch your breath because you feel something you've never experienced before—the floor is moving under your feet! At the same moment, your legs feel like rubber and you are aware of a rumbling sound.

What is happening? How long will it go on? Is something wrong with me? Am I losing it?

The hanging plants in the kitchen begin to sway and the juice glass on the table tips over and falls and shatters on the floor.

You are frozen with fascination and fright. Then the noise, like a horrible roar from a subterranean monster, assaults you full force. Your heart begins to pound wildly as the improbable idea of an earthquake begins to take hold.

Chaos suddenly erupts inside your house—kitchen cabinets open and rows of cups, plates and glasses fall to their death. Windows shake wildly, lamps fall over and light fixtures shatter. A bookcase tilts over, spilling its contents face-down.

Now the chemicals of panic shoot into your veins. Time, motion and stability dissolve into a psychedelic nightmare. Your perception hones in on the thousands of fascinating and unusual little events happening around you simultaneously.

Slowly, very slowly, a window develops miniscule cracks. First just a few, but then a fascinating configuration of scores, then hundreds, then seemingly thousands of intricate cracks which, in one instant, separate from each other. They are impossibly separated in mid-air, hanging there, and then slowly move out and away from each other. Beauty, fascination and horror. Each fragment is gliding through space, floating like a starship on an exploding intergalactic journey. But they're moving toward you, so you drop to the floor.

The glass fragments whiz by, their whooshes intermingling with the hundreds of other sounds of a house being demolished from within and without. WHEN WILL IT STOP!?

You watch as pictures drop to the floor and shatter, as furniture tips over and slides both gracefully and erratically along the floor. A hanging plant bounces off the ceiling and you grab for something solid and stable to hang onto.

Cracks ever so slowly appear in the walls, the ceiling, the floor. Will the house collapse? With one hand, you grab onto a table leg, but this once solid object feels liquid like the floor which is bucking and flowing beneath you.

Popping sounds! Get out of here, run! You can barely lift yourself up. The back door, really only a few feet away, seems miles distant. Its shape changes from a left parallelogram to a right parallelogram, out of sync with the doorframe which is changing in an opposite warp. It's hopeless!

Then you view the outside world and it's as insane as the inside. The road, driveway and lawn are undulating wildly. An evil asphalt serpent from hell is riding waves of grass down your street. STOP IT!!

You hear a shriek, then a shout, then screams and you're not sure if any of this came from you. You watch the neighborhood's pride—tall fir and pine trees—dancing wildly with each other, their tops whipsawing in agony. One giant falls, ever so gracefully and gently, straight onto your garage.

The noise! Electric wire flashes of light produce a crackling which blends into the sounds of boards crunching and creaking, groaning walls, water gurgling (from where?!). It's deafening.

Walls are bulging, swaying in and out, closing in on you. Everything is being pushed up and down and sideways all at once. A thin rivulet of blood winds along your right arm. Then you feel your house leave its foundation! You are in a free, storm-tossed boat, a little house of horrors in the twilight zone....So this is what the end of the world is like. Maybe you should have talked with all those Jehovah's Witnesses who came to your door after all!

You suddenly feel exhausted from the realization that you have no control over any of this. You must simply accept the cosmic humor of it all or go mad. You've been buffeted for what seems like hours. And at this moment, your emotions roller-coast to feelings of boredom. Every possibility of chaos has presented itself. There can be nothing new.

Except that it's slowing! And then, it's over. With shaky legs, you stagger toward what remains of the door. Something is on your hand. You look down to see you're still holding the coffee cup. And there's your watch which reads 8:00. Three minutes. Three minutes! It's not possible!

Outside is a wailing of car alarms and screaming. A huge cloud of dust rises over everything. Your house is a shambles. You are totally exhausted with fear and anger. With the last bit of energy you can muster, you shout, "WHY DIDN'T SOMEONE WARN US??!"

Well, consider yourself warned.

The two stories you've just read are geologically-based fiction. The first occurred in the late 1600's, many years before whites conquered what is now called the Washington coast, and the second will happen sometime in the perhaps-not-too-distant future to many people who live in Washington west of the Cascades.

Until the early 1980's, we were innocents as far as great earthquakes were concerned. Most of us had some knowledge of the San Francisco quake of 1906. Perhaps we had seen T.V. coverage of downed L.A. freeway overpasses in the 1970's. Some of us probably remember the unbelievable nightmare pictures of tangled destruction that came with the massive Alaskan quake of 1964. And older locals probably remember the Puget Sound tremors of 1949 and 1965.

Somehow, impossibly, Washington had been blessed by isolation. Big quakes happened much of the way from Chile to Alaska along the Pacific coast of the Americas. But the Northwest had been and always would be spared. After all, no one had *ever* experienced a giant tremor in Washington.

In a way, it didn't make sense, but in another way, it did. Although the basic ingredients for quakes were accepted and had been studied for a long time, these segments of the earth's crust, for whatever reason, had either stopped moving or were moving in such a way as to be harmless.

By the mid-1980's, however, the evidence began to accumulate that Washington had been visited by a great quake relatively recently and that such quakes occurred periodically.

Geologists began to uncover proof of major disruptions of soil which occurred every several hundred years or so on the coast and other areas of western Washington. Evidence accumulated that plant life and trees were buried and killed by salt water and landslides and that, in total, the evidence showed that similar occurences may have happened simultaneously throughout the state, from north to south and from the coast to the Cascades. All this dovetailed nicely with Native American legends that spoke of great earth and water movements.

The next two chapters will investigate how earthquakes happen and what recent evidence has been uncovered as to the probable magnitude and duration of the next big Washington jolt. Later, we'll look at the potential and hazards in the Puget Sound area, on the coast and along the Columbia River. We'll discuss how we can prepare individually and socially, so that we can best protect our lives and property. And we'll look at the role of government in disaster preparation and how we can make a difference.

So, stretch a bit, lie back and prepare to get tense!

CHAPTER 2

WHAT ARE THESE EARTHQUAKES AND WHY ARE THEY FOLLOWING ME?!

Let's get one thing straight from the beginning: I'm not *from* California, I just happened to stop through there for one and a half years on my journey to the Pacific Northwest from my home state of Texas. After this short stay, I was only too happy to leave the Golden State for a variety of reasons, not the least of which was that I had been fired from my job by then-Governor Ronald Reagan...but that's a story left for another time....

While in California I was, for the first time in my life, made aware of the real possibility of experiencing earthquakes. Although earthquake anxiety never kept me awake at night, an earthquake did once wake me up. One of the initial pluses, then, in leaving the state was the peace of mind in knowing that I would never again have to take earthquake hazards seriously. I was, unfortunately, an innocent in paradise. In preparing this handbook, I've discovered that earthquakes are a real possibility in most states in the U.S. Some regions of the country, in addition to California, appear to be potentially *very* dangerous—for example, portions of Alaska, Utah, Missouri, Montana, Nevada, South Carolina, Massachusetts and, gulp, much of the Pacific Northwest.

To comprehend our present danger, we must first understand what causes earthquakes. This chapter may be the most difficult in the book, but bear with me because you'll definitely benefit from this information.

Our problems begin deep below the floors of the world's oceans. As you probably know, the interior of the earth is extremely hot. The natural tendency of heat energy is to move to cooler places by the path of least resistance. The layer of our planet below the ocean floor is composed of hot soft rock which, when it moves toward the cooler outer crust, begins to melt, then oozes out as "magma" through cracks in the crust, creating underwater lava flows. Upon contact with the much cooler sea water, the lava hardens and attaches to the sides of the cracks.

This would end the story right here, except for the fact that more of the hot rock is heading upwards. According to one theory, its energy is always breaking through the earth's crust at these weakest seams. The result is that something must move out of the way and that something is the ocean floor on either side of the break. In these regions, then, new earth crust is in constant formation. *(See Figure 2.1, next page.)*

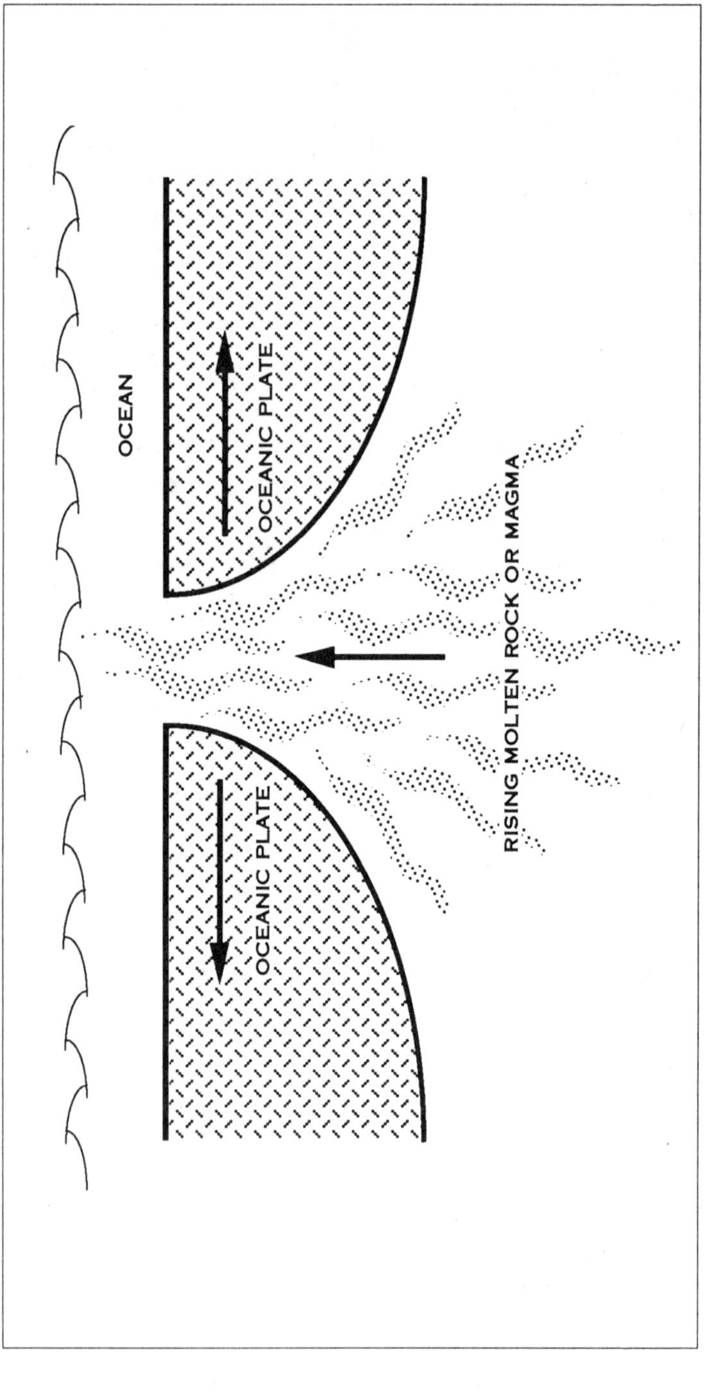

FIGURE 2.1. FORMATION OF THE EARTH'S CRUST

The earth's crust has, over hundreds of millions of years, developed many such breaks. The cracks have created huge slabs, or "plates," many miles thick, which are being pushed around on top of the layer just below, lubricated, much like a boat floats on water. Imagine a large number of boats bunched up next to each other in a small bay. If one boat is pushed, it will plow into or slide against one or two others. Unfortunately, this is what happens to the plates of the earth. Boundaries of plates, being pushed by expanding plate size, are always crunching up against or tightly sliding by other plate boundaries at the deceptively slow speed of one-half inch to several inches per year, the same rate as sea floor construction. The place where the boundaries come together is commonly called a "fault."

A very famous boundary in California is the San Andreas fault. Here two plates, the North American and the Pacific, slide past each other on land. *(See Figure 2.2, next page.)*

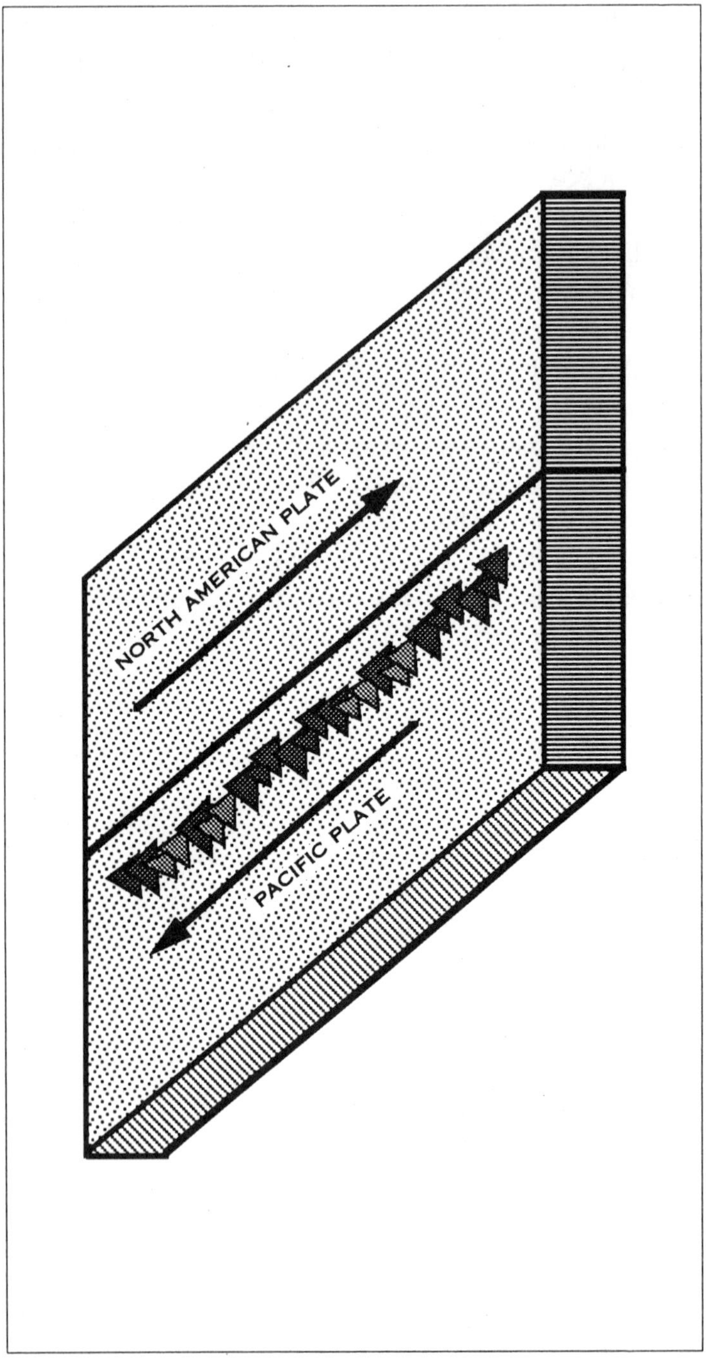

FIGURE 2.2. MOVEMENT OF THE SAN ANDREAS FAULT IN CALIFORNIA

Sometimes plates push straight against each other. If this happens on land, they plow together and buckle upward, forming spectacular mountain ranges. The Himalayas are an example of this type of movement. *(See Figure 2.3, below.)*

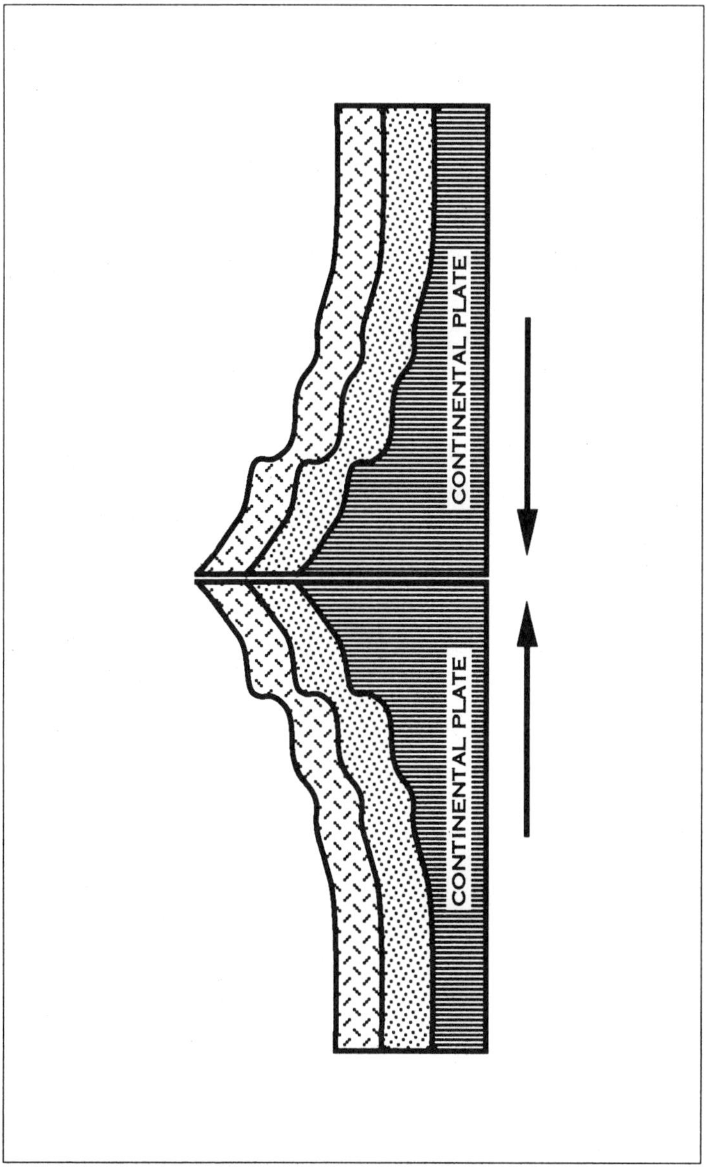

FIGURE 2.3. CONTINENTAL PLATE MOVEMENT CREATING MOUNTAIN RANGES

A third type of converging boundaries is more complex, more interesting and is the cause of the recurring great earthquakes in the Pacific Northwest. When a landmass, or "continental plate," meets a plate from the ocean, a dramatically different phenomenon occurs. For a variety of reasons, the ocean plate is thinner, lower and heavier than the land plate. As the two bump against each other just offshore, the continental plate rides over the oceanic one and the oceanic is pushed down toward the earth's interior. This type of convergence is called *subduction.* This is the crucial term for this handbook, a term Washingtonians will hear a lot in the future.

As pressure from newly-forming sea bed at one end pushes the oceanic plate farther down under the continental plate at the other end, the oceanic rock is subjected to great heat and pressure, which melts the top layer of this plate. However, being less dense than the continental rock above, it rises once again toward the surface in the form of magma (remember the process described earlier that starts the cycle). The recycling process has come full circle and has taken millions of years. This time, however, the process takes place on land and forms volcanic mountain ranges. These ranges are always relatively near the oceans and parallel the edges of landmasses. *(See Figure 2.4, next page.)*

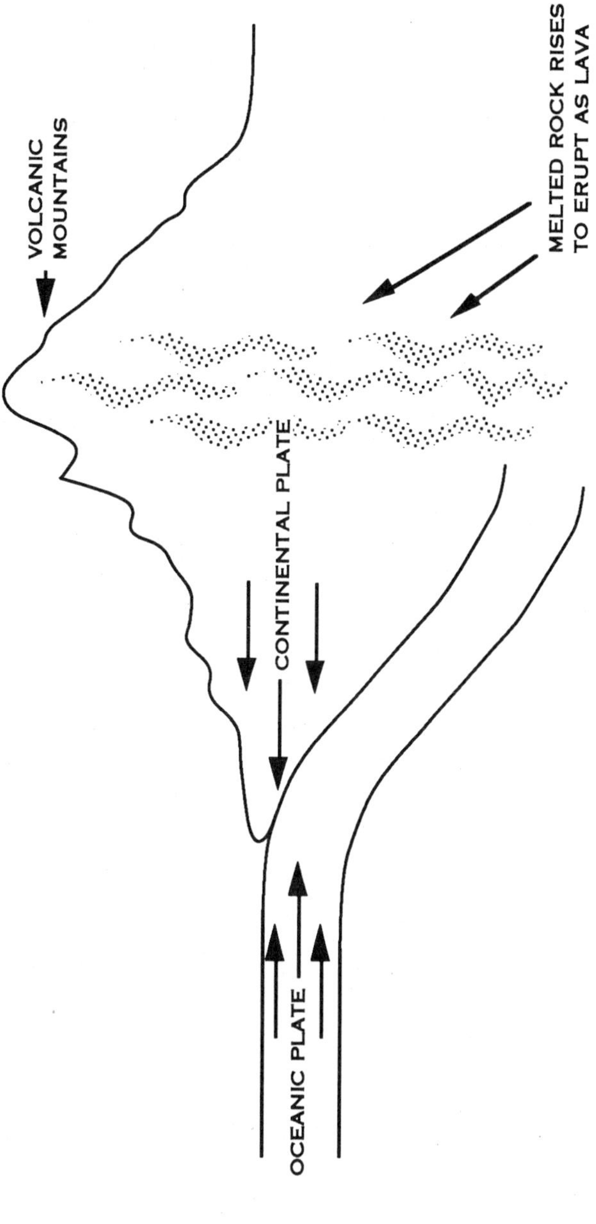

FIGURE 2.4. BASIC ELEMENTS OF SUBDUCTION

But what about earthquakes?

So far, we've described a geologic world in which, with the exception of head-on continental collisions, the earth's plates slide uneventfully past or under each other.

Were it only so!

We're all familiar with the dramatic earthquakes of the San Andreas fault in California. In this case, as we've said, two major plates slide past each other on land. If these plates did not stick together when they came into contact with each other or if they were smooth on their edges where they do touch, California would be an earthquake-safe area and housing there would be even more pricey than it is now!

Unfortunately, the plates are not smooth-edged, but rough. These jagged boundaries not only slide past, but also push up against the neighboring plates. Not too surprisingly, a section of one plate will often snag or hang up on the other. They are hooked and locked, but great forces are still relentlessly pushing them in opposite directions. Under this stress, they bend and strain, while, according to one theory, unbelievable amounts of energy are being stored up in the rocks.

Eventually, the strain exceeds the rocks' ability to withstand the stress and they break, allowing the plates to move to the places they would have been without the snagging. But the stored-up energy has been suddenly released and all hell breaks loose. *(See Figure 2.5, next page.)*

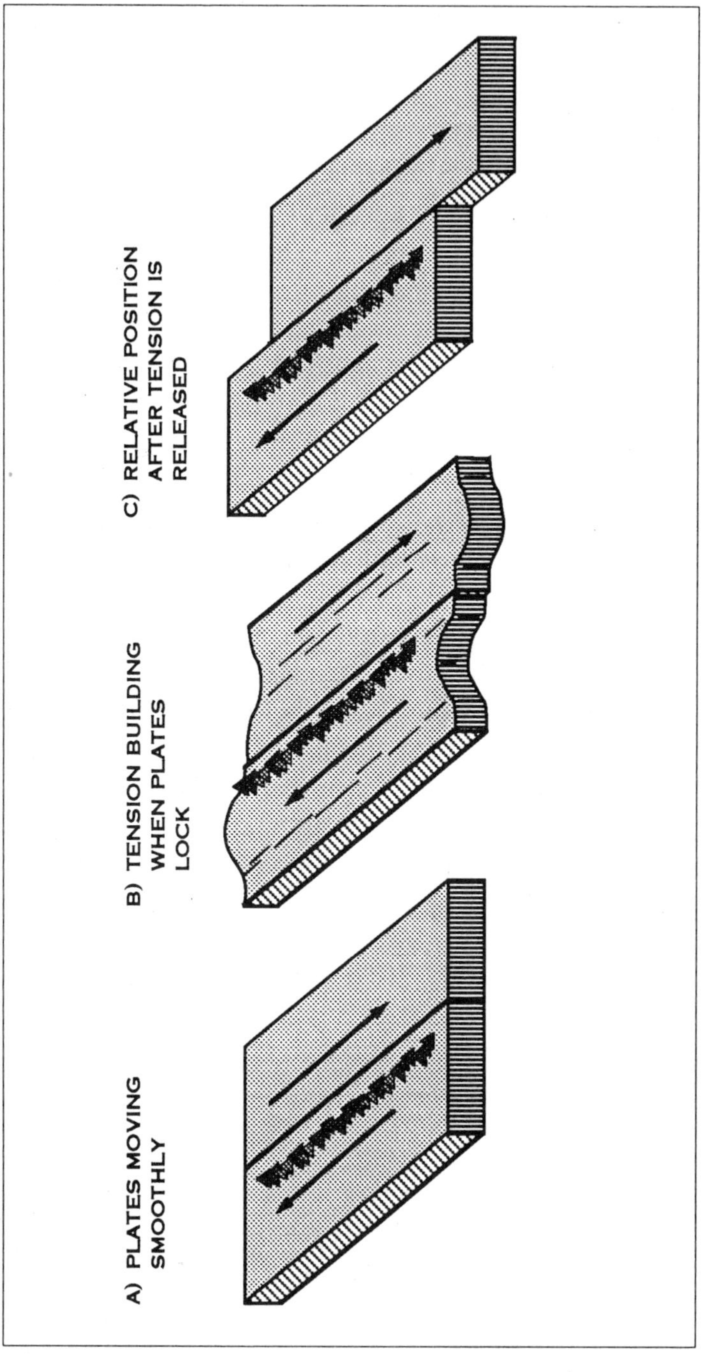

FIGURE 2.5. PLATE SLIPPAGE ON LAND RESULTING IN EARTHQUAKES

Although the strain leading up to a quake builds up over a very long period of time—decades or even centuries—the breakage for these "crustal quakes" usually lasts only a few seconds. The 1989 Loma Prieta "World Series" earthquake on the San Andreas fault shook the San Francisco Bay Area only ten to fifteen seconds. The main shock of the most famous North American quake, also in the Bay Area, in 1906, lasted less than a minute.

The most damaging quakes, however, are often the result of subduction zone breaks. The great tremors of 1985 in Mexico and of 1964 in Alaska are examples. These are the greatest magnitude quakes and their duration can be up to *five minutes*.

Just as with crustal quakes, plates in a subduction zone can also temporarily jam upon each other, not only at the point where the plates first meet, but at any point where the subducting plate makes contact with the upper plate. Once again, the pressure and stored energy can build up for centuries, with the pressure ultimately released as a great quake. Even without the plates locking, the process of subduction leads to stresses within the downgoing plate which can cause a quake. The sea floor plate descends as much as four hundred miles deep into the earth, where it is subject to intense compression, heat and stretching. *(See Figure 2.6, next page.)*

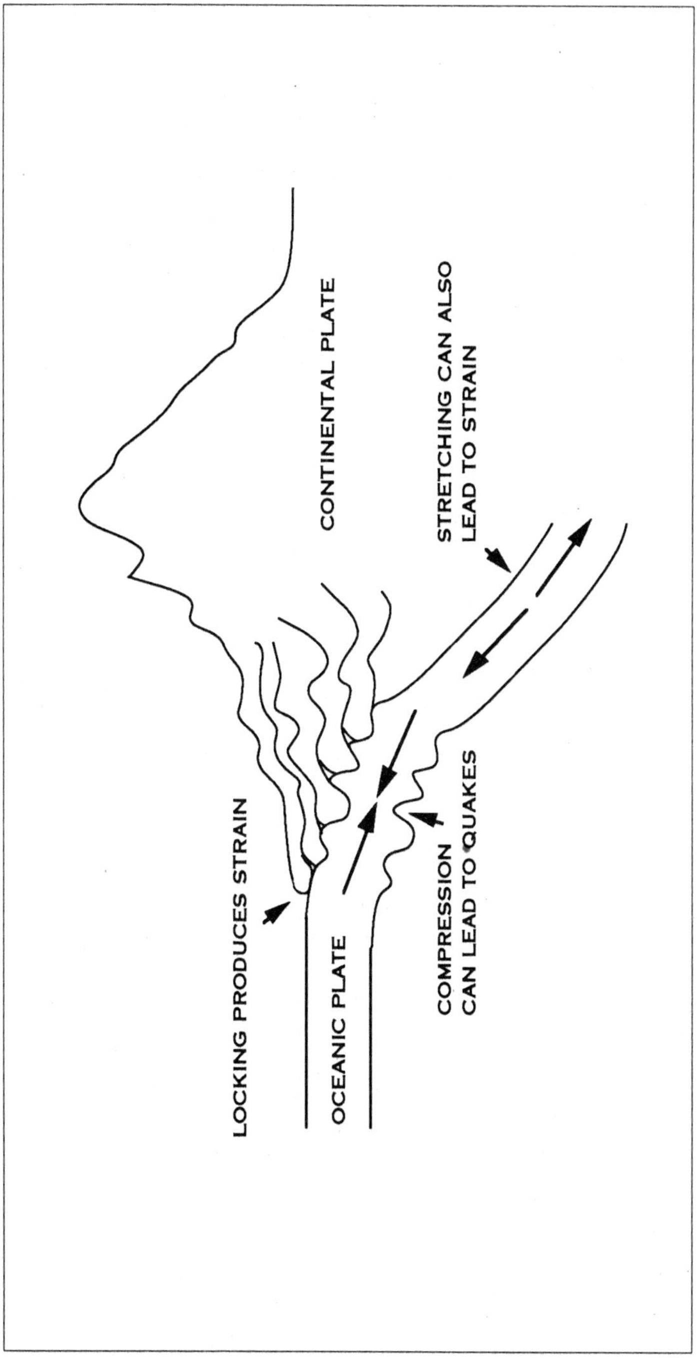

FIGURE 2.6. HOW SUBDUCTION CAN LEAD TO EARTHQUAKES

When a subduction earthquake occurs, the released energy causes the plates to move massively and rapidly. At their contact point in the ocean, one drops and the other rises. The energy transferred from the uplifted plate to the water above produces an action which is truly dramatic. The water molecules move back and forth with such force that their energy is passed on to adjacent molecules. This power then spreads out across the ocean or toward the nearby shoreline.

Called by the Japanese word "tsunami" (tsue-nah-me), this movement is not a "tidal wave"—it is not in any way caused by the tides and it is not only on the surface of the water. It is much more deadly and it can travel thousands of miles at speeds of hundreds of miles per hour. A tsunami created by the 1964 Alaska earthquake killed people and destroyed property as far away as the northern California coast. The physical properties of a tsunami allow it to be many miles long, but no more than two feet high in the open sea. However, when it beaches, it can be up to one hundred feet in height! In some instances, quakes can produce a series of deadly waves which all hit in a short period of time.

Finally in this chapter, a few words about measurement. We usually think of an earthquake in terms of a number—for example, the 1989 Bay Area quake was widely reported as a "7.1 on the Richter scale." (It was also known as the "Pretty Big One.") In addition, it was described in terms of the damage it did—a portion of an Oakland freeway collapsed; a section of the Bay Bridge fell; structures sustained heavy damage in Santa Cruz, Watsonville, the Marina District in San Francisco, and in other areas.

The Richter scale is an assessment of an earthquake using an instrument called a "seismograph" which measures the magnitude of the waves of energy that come from rock breakage. The advantage of this scale is that we can simply compare the numbers of different quakes to get an idea of their relative size and danger.

Unfortunately, there are lots of problems with Richter scale numbers. An earthquake of magnitude 7, for example, may cause extensive damage in one area, but because of factors we'll discuss extensively in later chapters (like building design, soil condition and distance from the epicenter), damage will be light in areas just a few miles away—or even just across the street! A Richter magnitude also doesn't let us know how far down in the ground the rock break occurred, how long the quake lasted or how much of the fault was involved.

Another approach to measuring a tremor, as we mentioned, is to describe and estimate damage to things and harm to people. This is called *intensity* and it is somewhat subjective. Two people, in the same spot and under stress, may see things quite differently. But a firsthand report can be much more illuminating than a seismograph reading.

CHAPTER 3

WELL, THIS SHOULD PUT US OFF THE MAP!

THE EVIDENCE FOR PAST AND FUTURE GREAT EARTHQUAKES IN WASHINGTON

It may come as a surprise to most Washingtonians to learn that the state ranks *second* in earthquake activity in the continental U.S. Look through books on physical geology and you'll see maps of the country which show historic seismicity displayed as dots which vary in size according to magnitude. Many states have a legacy of earthquakes—California, of course, but also Montana, Nevada, Idaho and Wyoming. Farther east such unlikely areas as Michigan, South Carolina, New York, Missouri and Massachusetts have had some fairly dangerous ones. It's safe to assume that states with lots of dots representing moderate to larger quakes will experience more in the future. And Washington has an impressive number of these dots.

But, until recently, geologists believed that Washington was safe from the truly giant subduction quakes which plague other parts of the world such as Alaska and Japan. The presence of a relatively small plate off the coast was known, but it was seismically quiet and the assumption was that it was characterized by smooth movement under the North American continental plate, if it was subducting at all. This was unusual since every similar zone in the world was the site of monster quakes. But there is such a thing as luck and there are always exceptions to the rule.

The plate in question, the Juan de Fuca oceanic plate, subducts along the coastline all the way from just south of the Oregon border to the middle of Vancouver Island, Canada. The plate is part of the area known as the Cascadia Subduction Zone, which includes the Gorda plate to the south and the Explorer plate to the north. At 625 miles in length, this subduction zone is *longer than* the San Andreas Fault. *(See Figure 3.1, below.)*

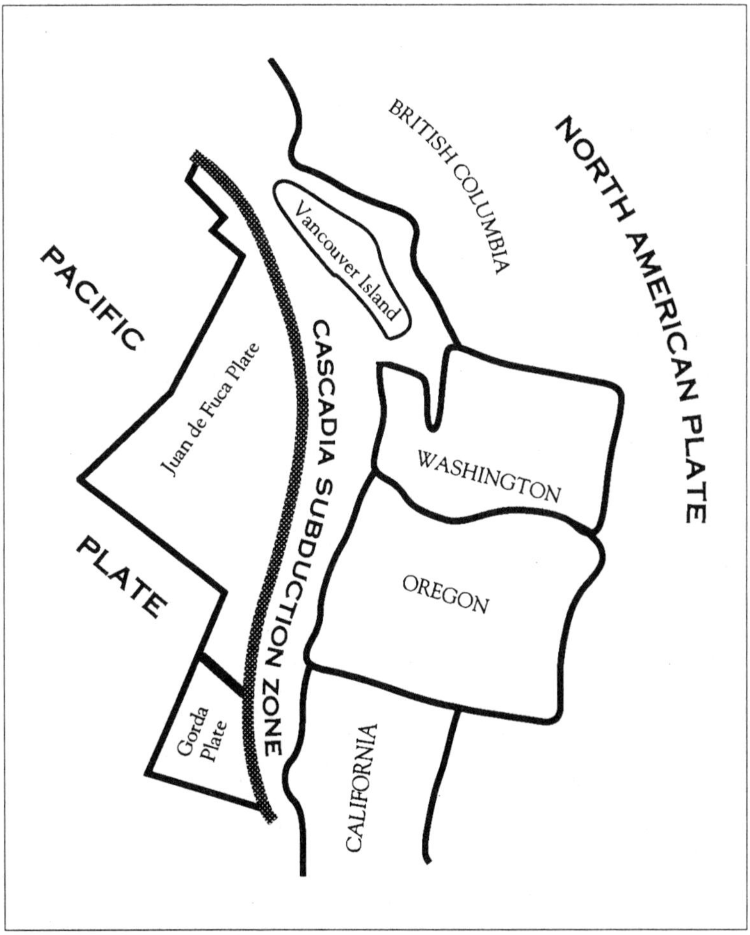

FIGURE 3.1. MAP OF THE CASCADIA SUBDUCTION ZONE

Other subduction zones off the coasts of southern Alaska, southern Mexico, Central America and off the whole Pacific coastline of South America down to central Chile were all identified as major producers of earthquakes.

But not ours! We're quiet, smooth, safe.

So, as late as the 1970's, the available evidence strongly suggested that, although subduction of the Juan de Fuca plate had occurred millions of years in the past, this subduction had ceased or even if it still continued, there was no evidence that it was developing in any way that could lead to future great quakes.

For example, volcanic activity appeared to be on the decline, which implied a slowing or a cessation of subduction. In 1980, Mt. St. Helens was to erupt and earlier, in 1914, Mt. Lassen, part of the Cascade chain, had exploded, but these events could be easily explained: As the top of the subducting plate heats and melts, as discussed earlier, it rises, melting the rock above and eventually emerges from volcanic mountains as lava; however, this process takes many thousands of years and these eruptions were perhaps caused by some of the last of the molten material slowly rising to the surface.

Similarly, although there was no significant seismic activity off the coast of the Pacific Northwest, quakes in the Puget Sound area—for example, in 1949 and 1965—had made it clear the region was earthquake active. The earlier quake, located between Olympia and Tacoma and measuring at 7.1 on the Richter scale, had killed eight people. The 1965 quake, centered between Tacoma and Seattle, led to the deaths of seven. Property damage was high in both. (If similar quakes were to hit the same areas today, the life and property toll could be much higher because of the development which has occurred in the intervening years.) Seismic instruments had determined that these quakes and many lesser ones in the Puget Sound area had occurred deep underneath the earth's crust and were most probably a result of stress in the subducted

Juan de Fuca plate. But since the quake activity was limited to this one area, it was argued, this was just the movement of one last plate fragment. Otherwise, movement appeared to have stopped.

However, at the same time, there was evidence for continuing subduction even without seismic activity. Scientific instruments seemed to show there was a tremendous amount of activity at the ridge several hundred miles out in the Pacific where underwater lava was supposedly forming the western boundary of the plate. This would surely mean that the plate was being pushed northeastward and under the North American plate.

Then, in 1983, research teams accumulated data which, when combined with earlier evidence, suggested that the Pacific Northwest might be in what is known as a "seismic gap." This term means that long fault lines are actually segmented, with each segment breaking in its own time, independent of the other segments. (This concept has recently received strong support from measurement of earthquake activity in California. The 1989 Bay Area quake involved a section of the northern portion of the San Andreas fault—the Loma Prieta segment—which had not slipped for a long time. Some segments, like the one that caused the 1906 San Francisco disaster, had already released their stresses, so the Loma Prieta segment was overdue. Scientists were so sure of this that earthquake alerts were issued for this segment just months before it broke!)

As mentioned earlier, there is an almost unbroken chain of recorded intense Pacific Coast subduction earthquake activity from Alaska to Chile. A seismic gap can display a lack of measurable activity because it is locked and therefore not moving, and seismic gaps do occur in subduction zones. Could it be, then, that the Cascadia Zone was quiet because it was stuck? *(See Figure 3.2, next page.)*

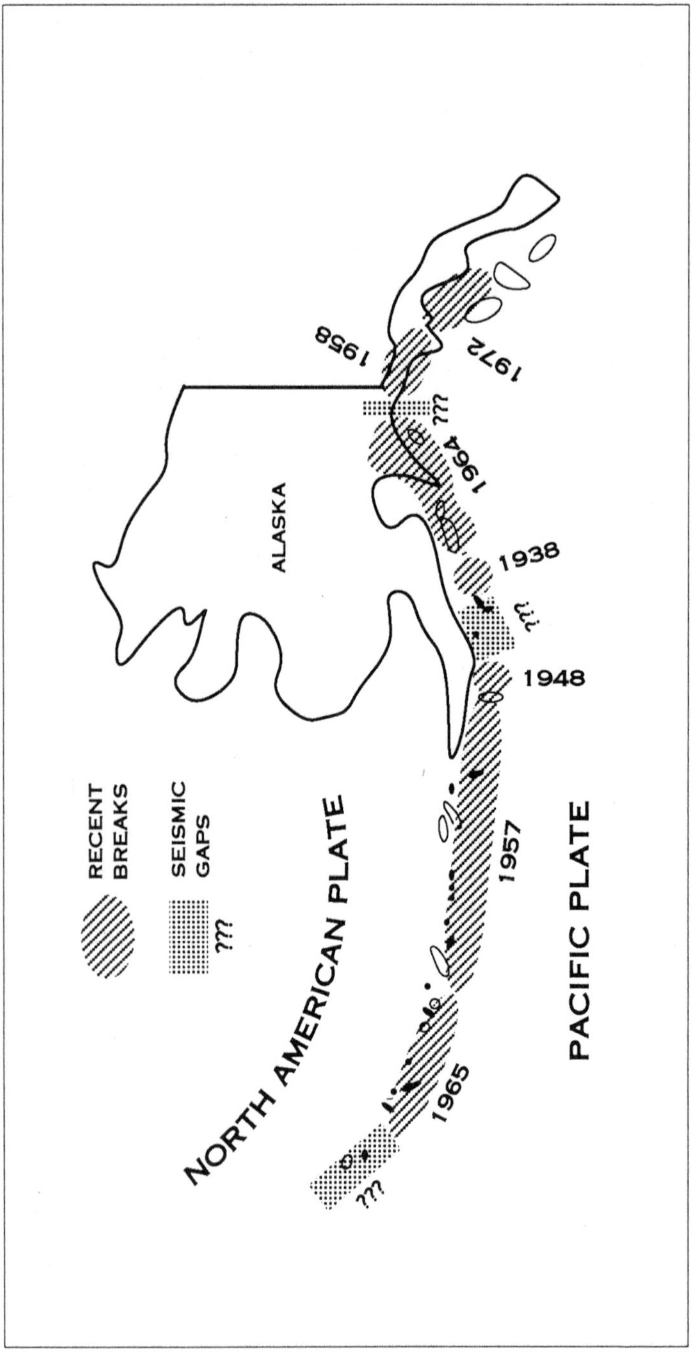

FIGURE 3.2. EXAMPLE OF FAULT LINE WITH SEISMIC GAPS

Another argument against massive earthquakes being produced by a subducting Juan de Fuca plate had to do with its youth. We're talking youth in *geologic* terms—the plate is about ten million years old! According to some scientists, a young plate would have little chance of snagging on an overriding plate because it is still relatively warm and pliable and so would simply flow past a potential problem. But in the mid-1980's, researchers produced evidence that other young subduction plates off the coasts of Chile, Colombia and Japan had been responsible for the highest magnitude quakes in recent history. Younger plates, they reasoned, are lighter and their descent is not as steep as older ones. This means they will probably have more contact with, and therefore have more potential for getting locked on, the upper plates.

The problem with all the evidence presented up to this point is that it's really just conjecture—it can't be seen or touched. With some faults that are on land, such as the San Andreas, geologists can both touch and see the solid evidence of past quakes because these faults break through to the surface of the ground or are at least close to the surface. However, because the Cascadia Subduction Zone lies under the ocean, what was lacking was hard physical evidence of recent prehistoric great quakes. There were theories, but not much physical fact.

Until, that is, the great quakes in Alaska in 1964 and in Chile in 1960. Comprehensive studies of these two events concluded that the locking of converging plates causes the overriding continental plate to rise up and then to collapse with the release of the strain. When this plate falls, many coastal areas are inundated with sea water for a period of time. The submerged land plants and trees are killed by this action and leave a record as a unique layer of soil or "sedimentation." In addition, remains of plants which grow only in salt water are deposited just above.

Armed with this information, Brian Atwater, a Seattle-based geologist, began in 1986 to dig marshy areas along the Washington coast to check for great quake-related sedimentation. His findings were dramatic and revolutionary. He was able to prove that points along the coast, from the tip of the Olympic Peninsula to the mouth of the Columbia River, had dropped, perhaps simultaneously, at least several times in relatively recent prehistoric periods, that the areas had also been hit by tsunamis and that no other explanations could as adequately explain his observations. The mud, as it were, had hit the fan.

The drops along the coast were not on a regular basis. Recurrence times between episodes appeared to vary from several hundred to one thousand years. These findings spurred research along the Oregon coast and the evidence there paralleled the Washington findings. So far, fifteen sites in Oregon between Seaside and the Coquille River have almost identical marsh records. Meanwhile, other investigators found evidence that large prehistoric landslides of underwater ridges along the Cascadia Zone had occurred, perhaps caused by large tremors.

Using data from coastal trees which had been partially submerged and killed by sea water, researchers in 1990 and 1991 published findings which estimated that the last major subduction quake occurred on the Washington coast, most probably in the mid-1680's.

Then, in April 1992, a large quake occurred in Northern California, centered near Petrolia. It has recently been determined that it was a Gorda plate subduction quake, the first verified subduction quake off the continental U.S. in recorded history.

Finally, in June 1993, scientific instruments detected volcanic eruptions and lava flows at the westernmost edge of the Juan de Fuca plate, definite proof that the plate mass is always growing and pushing toward the Pacific Northwest.

As a result of all this research, the generally accepted

conclusion among geologists today is that the probability of future great subduction quakes in the Pacific Northwest is quite high.

Okay, so WHEN will it hit, how BIG will it be, how LONG will it last, and WHERE will it happen??!!

With good reason, earthquake *prediction* is the major concern of the general public, the media and a growing body of scientific investigation. We all like to have a feeling of control over our lives. Since we have absolutely *no* control over the next superquake, the next best thing is to be prepared, and prediction is part of preparation. Without some feeling of knowledge and preparedness, we feel helpless.

Unfortunately, prediction of earthquakes is very difficult, if not impossible. First, the behavior of faults can vary quite a bit. The way a subduction zone acts before one quake may be very different from the way it behaves before another. Also, no two faults are exactly alike, so if we study one subduction fault a lot, we may not necessarily understand another fault like the Juan de Fuca much better. And, since most quakes occur fairly deep in the earth, it's very difficult to directly measure movement and locking in the crucial regions of the plates.

Having said all this, it is reassuring that *something* is being done. Like all modern sciences, geology has developed some rather impressive technology. Earthquakes can be measured not only while they're happening, but also while they're building up. An impressive array of instruments has been developed to measure changes in the earth that may precede tremors. One device measures surface slip on a fault. Another gauges levels of helium in the soil because there's some evidence that detectable amounts of this gas increase in the days and hours before big tremors. Another meter detects rock deformation that may precede a quake. Devices which monitor very low frequency radio waves may also hold some promise. Instruments in the Santa Cruz mountains in California went wild with activity just hours before the Loma Prieta quake; another reacted strongly before a major Japanese tremor in 1989.

As we mentioned earlier, seismographs gauge movement of the earth's surface. Not only do they measure the primary shocks, but they can also detect early slippage before the big break, just like a branch slowly cracks before it snaps. These "foreshocks" are sometimes detected, but unfortunately, they are usually only identified as such after the fact because they look like other movements which do not precede earthquakes. Still, there is hope that advances in this technology combined with better data analysis will, in the future, allow scientists to be able to issue adequate warnings.

Finally, we mention two quite different techniques being used to measure land movement caused by strain along the Cascadia Subduction Zone. The first involves several years of research on the Olympic Peninsula, on Puget Sound and in the Astoria, Oregon, area. Using very sophisticated laser technology, scientists carefully measure tiny amounts of soil uplift on a daily basis. The second involves highways along the Oregon coast which are resurfaced on a regular basis. Survey crews simply note how much the land has bent upwards since the last resurfacing. Although the changes noted with these techniques are small each year, they are nonetheless significant in comparison with the amount of warping in other areas which are not affected by subduction strain.

Once again: where, when, how big and for how long?!

Because the oceanic section of the Juan de Fuca plate extends the length of the Oregon-Washington coastline and since we also now know that the subducted portion of this plate is under all of western Washington east to the Cascade range, *any* area of western Washington can be affected by subduction quakes. The length and position of the sections of the zone which rupture will determine which ground areas will be shaken.

"How big"—that is, the magnitude—will also be influenced by how much of the plate snaps. If a very large segment or all of the plate breaks at once, the magnitude will be enormous, for example, up to a 9. This is near the upper limit of possible earthquake size and only several this large have occurred worldwide in the last century. Another possibility is that slippage would occur as several great quakes (magnitude 8+ each) over a time span of up to ten years. One geologist calls this possibility the "Decade of Terror."

The intensity—that is, the amount of shaking experienced and the resulting damage—is likewise determined by how much of the fault is involved in the break. As mentioned earlier, intensity is also affected by the type of soil and rock we are on. In addition, the intensity of the tremor is influenced by the duration and, in general, the larger the break, the longer the shaking time. The worst case scenario for the Pacific Northwest would be similar to the four minutes experienced by Alaskans in 1964.

Now the most important question of all: *when* will the next one hit? To some extent, the answer to this relates to how often great quakes happen in Washington and when the last one occurred. Unfortunately, the evidence accumulated to this point does not necessarily lead to a conclusive answer and there may not be a good answer until a lot more research has been conducted. Estimates from the scientific community are simply educated guesses. According to some research, the time between great quakes has *averaged* about three hundred and fifty years. But average recurrence is misleading since the actual time between great quakes has ranged from as little as perhaps one hundred and fifty years to many hundreds of years.

However, some recent, sobering research suggests that coastal land may have already risen to the minimum height required for the fault to break. And sophisticated measures of compression of the earth in the Puget Sound–

Seattle area also suggest stress between the plates may have built to a breaking point. Thus, we are perhaps in a "window" where a great quake could happen at anytime from now until some decades from now.

Since this time frame could be a long span of time, not just a year or several years, should we just say: let the next generation worry about it, the likelihood is greater for them? Should we panic, leave the state and move in with Auntie Em in Kansas? Or should we decide to remain in this wonderland and ride it out? Since there's no history of significant migration from earthquake-dangerous areas anywhere in the world, the vast majority of us will stay put. In fact, population and property values soared during the 1970's and 1980's in areas long known to be very dangerous, such as Anchorage, Tokyo and Los Angeles. And when property values recently did come down in these areas, it wasn't because of earthquake fears.

So let's pull out the old clichés—better safe than sorry—better to err on the side of caution—and start preparing ourselves.

The remainder of this handbook will be concerned with the specifics of earthquake hazards and how to successfully deal with them.

PART TWO

GETTING UNSTUCK:
MAYBE IT IS OUR FAULT AND WE SHOULD DO SOMETHING ABOUT IT

THE SOLUTION

CHAPTER 4

THE COMING COLLAPSE AND HOW TO SURVIVE IT:

PERSONAL, FAMILY AND HOME SAFETY

It seems a good bet that most Washingtonians have no stored supplies for an emergency and know little about good survival techniques for during and after a major disaster.

We're not talking about survivalist methods—that is, stocking lots of food, weapons and camouflage gear so as to protect our food, water and possessions *from* others who might want them. In fact, after major disasters, most people *help each other* in ways they wouldn't have before. Chapters 5 and 6 will deal with ways communities can plan to handle a great earthquake; but to ensure that community response is successful, we must also promote individual preparedness which can tide us over until major relief efforts kick in. So, individually and as communities, we need to keep a lot of life-sustaining essentials in storage.

First imagine what a great quake could do to your house or apartment, workplace, and to the public and private services we take for granted. Depending on how well your dwelling is built, the types of materials used in its construction, the soil beneath it, and the magnitude of the earthquake, you may well find yourself, for all practical purposes, homeless for a while.

Walls and roofs can collapse, frames can slide off foundations. Or the structure might be so weakened that an aftershock could cause more damage than the primary shock. It may simply be too dangerous to consider remaining in your home or even going back in to get needed supplies. Depending on the size of a disaster, it could be a month or two—that's right, a month or two—before electricity and water service are restored. Gas and sewer lines may be broken and phone service may be down. Car travel may be impossible due to damaged roadways and debris which could stop relief vehicles. It may be cold or wet or windy or dark or all of these.

We will need liquids, food, lighting, protection from the elements and, perhaps, first aid. So before the disaster hits, we need to set aside a lot of emergency supplies in easily accessible and protected areas. The availability of such areas and the amount we can afford to put out for this will vary from person to person. If you live alone, your needs will be somewhat different than if you're in a family. The lists and rationales for these lists that follow are only guidelines and suggestions. It's up to you to tailor them to your own situation. What will *you* need to survive adequately for one to several weeks? To be on the safe side, it's best to prepare for the worst possible event. Remember also that some of the supplies you put aside for use after an earthquake can also be used for other emergencies such as ice storms, heavy snows, fires or windstorms.

THE BASIC HOME EARTHQUAKE EMERGENCY KIT

(Use these lists as a checklist.)

☛ Flashlights (one per person) and/or portable battery-operated lanterns. Alkaline batteries should be used because they last longer and don't leak. Keep all batteries—and extras—in the packages you buy them in or in the flashlight or lantern. Replace batteries after one year.

☛ <u>Portable radio with alkaline batteries and spares.</u> Try not to use your car radio as it drains the battery.

☛ <u>Needle and thread</u> (scissors should be kept in first aid kit).

☛ <u>First Aid Kit</u>

- ❒ First aid manual
- ❒ Extra pair of prescription glasses
- ❒ Necessary prescription drugs, medications, etc.
- ❒ Aspirin or substitute
- ❒ Antibiotic ointment
- ❒ Rubbing alcohol
- ❒ Tincture of iodine
- ❒ Anti-diarrhea medication
- ❒ Scissors
- ❒ Tweezers
- ❒ Safety pins
- ❒ Sterile latex gloves
- ❒ Thermometer
- ❒ Petroleum jelly
- ❒ Tissues
- ❒ Cotton-tipped swabs
- ❒ Sterile absorbent cotton
- ❒ Box of bandaids
- ❒ Ace bandages
- ❒ Triangle bandages
- ❒ Butterfly bandages
- ❒ Adhesive tape
- ❒ Sterile gauze roll bandages

- ❒ Gauze pads
- ❒ Pocket mask with one-way valve
- ❒ Whistle

It's best to keep all this in one tightly-sealing box marked "First Aid"—a fishing tackle box is good.

Prescription glasses can be extras or ones that you've "outgrown," but are still adequate.

Any liquids and glass bottles should be sealed in zip-shut storage bags.

Check expiration dates and make notes when to replace.

Take a first aid and a CPR class or refresher course—hospitals, clinics and other medical facilities may be damaged and overloaded. Your skills might be the difference between life and death. A first aid manual should be kept in the kit for more complete information. If you don't already have one, we highly recommend the manuals available from your local Red Cross Chapter. You might also be able to pick up a pocket mask which gives you protection while giving rescue breathing to a stranger.

☛ <u>Food.</u> For example:

- ❒ Freeze-dried food
- ❒ High protein and granola bars
- ❒ Nuts (dry roasted and unsalted)
- ❒ Canned fish
- ❒ Canned fruit and juices
- ❒ Dried fruit
- ❒ Canned and packaged soup
- ❒ Canned chili, beef stew, chicken, etc.

- ❒ Canned spaghetti
- ❒ Canned beans
- ❒ Hot cereals (sealed in zip-shut bag)
- ❒ Peanut butter
- ❒ Powdered and canned milk
- ❒ Honey
- ❒ Trail mix

The most important thing is to have a wide variety of foods that you like. Life will be grueling enough without having to deal with dull food. The less cooking you have to do and the easier it is to heat, the better. It's best to have food for one to two weeks. Restock the canned foods every twelve months. (Most of us have enough food in our houses to last at least several days.) If electricity is off, frozen food lasts about three days in an unopened freezer. *But note: after only two hours at room temperature, most perishable foods are dangerous to eat!* There are also companies that package more convenient (and also more expensive) ration kits designed specifically for disaster preparedness.

☛ Clothes and Weather Protection

- ❒ Jackets
- ❒ Hats
- ❒ Gloves
- ❒ Changes of underwear, insulated underwear
- ❒ Socks
- ❒ Handkerchiefs
- ❒ Pants
- ❒ Shirts, blouses

- ❒ Thick-soled shoes
- ❒ Blankets
- ❒ Plastic rain gear
- ❒ Tent
- ❒ Tarp
- ❒ Sleeping bag and insulated sleeping pad

Clothes should be comfortable. It's important to stay warm and dry and clean. Thick-soled shoes are necessary because of broken glass and other sharp objects and downed wires. Gloves should be warm, but also heavy-duty, as digging through rubble could be dangerous to your hands. You might also buy foil "space" blankets not only for your own warmth, but also to wrap around anyone who is in shock. Your tent can be set up in your yard unless it would be close to utility poles, trees or a structure that could fall from aftershocks. Setting up the tent in a park or a school yard or any open area might be safer if you're willing to leave your home and possessions. The tarp is to cover the impressive array of survival material!

☛ <u>Other Camping Gear</u>

- ❒ Portable cooking device
- ❒ Sterno® canned heat
- ❒ Saucepan with lid
- ❒ Candles
- ❒ Waterproof matches, butane lighter or a halizone non-extinguishable lighter
- ❒ Reusable plastic plates, cups, utensils
- ❒ Utility knife with can opener
- ❒ 50-foot nylon utility cord
- ❒ Bungee cords

Great care should be used with matches and all flames if there is a possibility of gas leaks. This danger exists wherever gas lines are present—in or near houses and buildings, in parks, under roads, etc. <u>If you smell gas, don't use anything that produces a spark—for example, matches, flints or cigarette lighters!</u> Also remember, aftershocks can topple candles.

☛ <u>Tools and Related Supplies</u>

- ❒ Adjustable, 12-inch crescent wrench
- ❒ Hammer
- ❒ Nails
- ❒ Pliers
- ❒ Screwdrivers (Phillips and flat blade)
- ❒ Crowbar
- ❒ Hatchet
- ❒ Duct tape
- ❒ Plastic sheeting

The crescent wrench may be needed for turning off the main gas valve. Also, check to see if you need a special tool for turning off your main water line. The other tools are primarily for rescue work and for securing and sealing damaged areas of your home.

☛ <u>Water and Sanitation</u>

- ❒ 1/2 to 1 gallon of water per person per day
- ❒ Shovel
- ❒ Bucket or small trash can with tight-fitting lid or portable toilet
- ❒ Large zip-shut plastic bags
- ❒ Small and large plastic garbage bags with twist ties

- ❐ Toothbrush
- ❐ Shampoo
- ❐ Comb or brush
- ❐ Deodorant
- ❐ Hand soap
- ❐ Face cloth
- ❐ Bath towels
- ❐ Detergent
- ❐ Dish pan
- ❐ Sanitary napkins
- ❐ Toilet paper
- ❐ Bag of chlorinated lime powder or a household disinfectant
- ❐ Liquid (<u>not granular!</u>) chlorine bleach in small plastic bottles (<u>must be</u> 5.25% sodium hypochlorite)

  or

  Iodine or chlorine tablets
- ❐ Eyedropper (tape to bleach or iodine bottle and <u>only</u> use for either bleach or iodine)
- ❐ Shaving articles

If water lines are broken, stored water will have to suffice until large-scale relief efforts are underway. Water may be the most important item in the disaster kit. Humans cannot survive for more than a few days without water! Store water separate from your other supplies. <u>For each person</u>, we keep five gallons of pure drinking water and five gallons for washing.

The drinking water is the commercially-bottled type in 2.5 gallon plastic containers. If kept in a dark, cool, dry place, bottled water can keep for one year. The washing water is kept in five-gallon plastic utility containers, also in a cool, dry place and should be changed once a month.

Unless chemicals have been added, water from your toilet tank (not bowl!!) is a safe source for about seven gallons of drinking water(but purify first, according to directions below). You can also get water from canned goods and melting ice cubes.

Water drained from an undamaged water heater (40 to 80 gallons) or from creeks, streams, etc., can be made safe by first straining through a cloth and then boiling for a few minutes or by using the iodine or chlorine tablets (follow directions on the bottle). Replace the tablets every two years.

To purify with liquid bleach, add 2 drops with the eyedropper to each quart of water (4 drops if water is cloudy), mix well and let stand for 30 minutes. If there's no chlorine smell to the water, repeat the process. Do not use granular bleach, as it is poisonous! Replace the liquid bleach each year, as it loses its potency

A 2% tincture of iodine can also be used. Use 3 drops for each quart of water, 6 drops to cloudy water.

If sewer and/or water lines are broken, you should not flush the toilet and you must find a safe way to dispose of human waste. The easiest approach, of course, is to have a chemical toilet on hand. Or you can use your shovel and dig an "outhouse" with a large plastic bag opened in the hole. The bag can be changed when necessary. Or forget the bag and just throw in some lime each time the "toilet" is used. (Lime can be purchased at a building supply store.) Or you can line a bucket with a smaller plastic bag and dispose of it far from humans or bury it. The large plastic bags can also be used for trash. All bags should be closed tightly with twist ties.

Finally, although shaving might appear to be a nonessential luxury, it's best to maintain a normal routine as much as possible in order to counter the stressful situation.

☛ <u>A Few More Necessary Supplies</u>

- ❒ Coins
- ❒ Cash
- ❒ Large notebook
- ❒ Pencil, pen
- ❒ Postcards
- ❒ Stamps
- ❒ Car and house keys
- ❒ Reading materials for all family members
- ❒ Copies of important papers and documents (keep the originals in a bank safe deposit box)
- ❒ Pet food
- ❒ Baby supplies—diapers, formula, wipes, etc., etc.
- ❒ Toys, portable games, deck of cards
- ❒ A local map
- ❒ This handbook!

The coins are for pay phone emergency calls in case your phone is dead. The cash is for buying food, gas, a motel room, etc. The notebook should have important addresses and phone numbers, bank account numbers and the like, plus a room-by-room listing of all your possessions which are covered by your earthquake insurance *(see pages 82 to 84)*. You can also keep an earthquake diary which will become an interesting part of the family history for your heirs.

The mail should go through even with large disasters. So send postcards to distant friends and relatives or even folks across town to let them know how you are. (Take it from an antiques dealer—these will also be <u>very</u> desirable collectibles in the future!) Or if you have friends and relatives who live outside the Pacific Northwest, you should choose one of these as a contact person you can get in touch with after the quake. A postcard or letter will do if long-distance phone service is down or not accessible. Because it's easier to call out of a disaster area than to call into it, your other friends can call this contact to find out how you are. Obviously, let them all know who the contact is as soon as possible—don't wait till the earthquake hits.

The local map will help you decide how best to get to the nearest relief shelter if you're told to leave your neighborhood. This is especially important if damaged or blocked roads or downed power lines force you to find an alternate, unfamiliar route.

☛ **Fire extinguishers:** A-B-C type. The more the merrier. Mount on wall studs so they won't shake off. Keep in the kitchen, the bedroom and the garage. The A-B-C type can be used for all types of fires.

In addition, keep a bucket of sand next to your wood stove or fireplace for the sobering possibility that the quake hits while you're heating with wood. If water mains are not broken, use your garden hoses to deal with smaller fires.

☛ **Mini Kits** It would also be a smart idea to have a box under your bed which contains a flashlight, thick-soled shoes or slippers, work gloves, glasses, house and car keys, and some clothes you can get into fast in case you're asleep when a tremor hits. If you work some distance from your home, it might also be a good idea to keep some bottled water, a change of clothes and three days' supply of food at your workplace. In addition, a mini kit in each car would come in handy not only for natural disasters, but also for accidents and mechanical problems. Include a flashlight, a small first aid kit, a warm coat, one gallon of bottled water, a fire extinguisher and a copy of this handbook.

NOTE: If you live on the coast, have a backpack for each member of the family which contains warm clothes, rain gear, bottled water and dried food, a flashlight, and this handbook. After a quake, you will have to get to high ground as soon as possible. There is a good possibility that you will have to walk or run there since roads may not be passable.

This list of supplies may seem mind-boggling. After all, *where* can you store *all this stuff?!* The ideal area might be outside your home; for example, in an attached or separate garage or in a sturdy out-building. (Free-standing storage units which can withstand earthquakes are manufactured in the Pacific Northwest.) If you don't have either of these or if you live in an apartment, use a closet with a good latch near your front door.

Also, what type of containers should be used? One possibility which we're using at present is a large garbage can with wheels and a lid that locks shut. It's mobile and can haul a lot of weight easily. Pack the can so that the flashlight and radio are on top, with first aid material right underneath. The next layer can be food, with clothes and other supplies at the bottom. We also have a backpack filled with essential materials for a quick getaway. Heavy cardboard cartons, small suitcases and dufflebags are other good choices.

One note of caution: keep water supplies separate. We found this out the hard way several years ago. We had last opened our garbage can which contains much of our disaster supplies about six month earlier. In the interim, a one-gallon water bottle had leaked all its contents and mold had destroyed almost everything underneath.

Which brings up the cost of stocking an earthquake kit. To minimize costs, we've bought a lot of our supplies at thrift stores and garage and estate sales. Also, check department and sporting goods stores during sales.

Finally, always keep your car's gas tank at least half full. Gasoline may not be available for quite a while after a quake. You may be ordered to evacuate your area, so you'll want to be able to haul all your supplies to a safer area.

MAKING YOUR HOME EARTHQUAKE SAFER

1) Securing Articles in the Home

Just about every object in your home could, depending on the intensity of an earthquake, fall, move, fly, spill or break. As in the previous section on supplies, we're listing a large number of ways to secure possessions. Once again, you can pick and choose—you have to decide just how much time, money, energy, and inconvenience you feel is necessary for your own circumstances and temperament. You can go to extreme lengths to protect *everything* or you can just secure a few crucial objects. Also, keep in mind that even though you can earthquake-insure all your possessions, there will be a deductible and there's still the chore of replacing all the items which are damaged or destroyed.

As of this writing, buying all the materials you need to secure your home's interior is a hassle and means going to a variety of stores. (We've never heard of any store in Washington advertise that it sells "earthquake supplies.") A hardware, lumber supply or home improvement store will be a good first stop. You'll probably have to check out specialty stores, too; for example, those which deal just in computer or stereo supplies. Walk the aisles with a list and get advice from the salespeople. Mention what you want to do and why you're doing it. Eventually, stores in Washington will realize there's a market for earthquake preparedness. In California, there are stores which only sell disaster preparedness supplies! Perhaps, in time, we'll also have stores supported by government funding which will buy materials in quantity and sell at or near cost, with profits going to programs to earthquake-protect houses at no cost for low-income, elderly and handicapped citizens.

Now let's again imagine what the contents of your home will do in a great quake. Closets and cupboard doors will fly open and things in them will spill out. If they're breakable, they'll probably break. If they have liquid in them, they may spill. Objects hanging on walls and ceilings may fall and smash to pieces. Free-standing heaters, lamps, tall furniture, and bookcases could topple. Appliances and other heavy objects could begin a very dangerous walking tour of your home! So take a slow walk around (if you have children, make it a game and have them help) and imagine what would happen if your rooms were bouncing around. Take notes, decide what's most important, what actions will make a lot of difference for a small outlay of time and money, and do those things first.

❒ **The Water Heater**

If not secured, this appliance might fall over and spill its contents, not only making a mess and necessitating buying another one, but also depriving you of many gallons of much-needed water. It could also snap its gas or electric line and start a fire or short out. Sounds like a water heater is a good first project! In addition to being perhaps the one most important item to secure, modifications are fairly cheap and easy and won't cramp your style in the slightest. *(See Figure 4.1, next page.)*

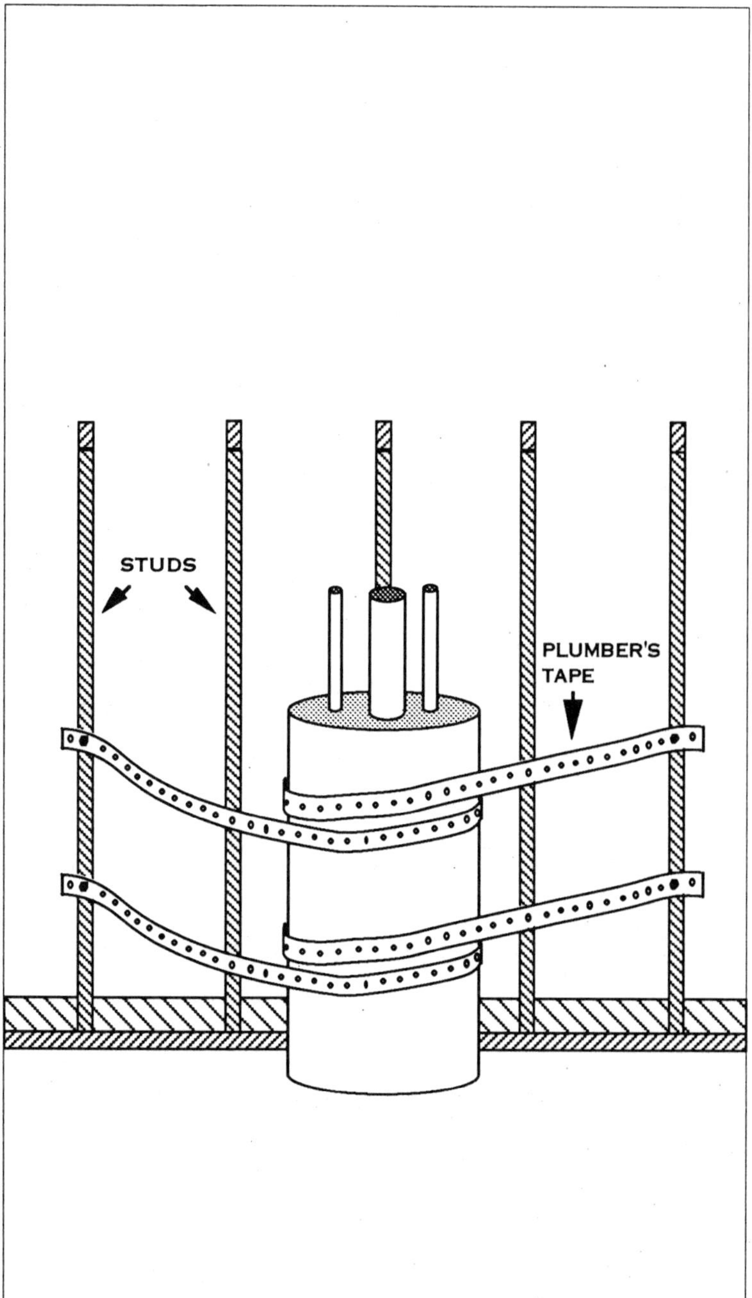

FIGURE 4.1. SECURED WATER HEATER

As you can see, there are a couple of ways to safeguard this very important appliance. First, if there are holes at the base for bolting the heater to the floor, bolt it. Whether this is the case or not, you should also secure the water heater with earthquake strapping or two strips of plumber's tape (metal, with perforations), as shown. To measure the lengths you'll need, first find the wall studs on either side of the heater. If you have drywall, you can use a "stud finder," which can be purchased at a building supply store. Tack one end of a cloth tape measure into the wall at one stud, wrap the tape once around the heater and then bring it to the other stud. Add several inches per strip just to be sure. One strip should be about six inches from the top and the other about one and a half feet from the bottom. Bolt the ends of the plumber's tape or earthquake strapping into the studs with 3" lag bolts with washers. Make sure the fit is tight.

Another possibility is to purchase a special water heater earthquake bracket from a hardware store and install it to the top of the heater.

While you're at it, also strap (and bolt down, if possible) all free-standing heating units such as wood stoves and propane tanks. Gas lines should be flexible to allow some room to move without breaking.

Because you may draw drinking water from the water heater in an emergency, it's a good idea to partially drain it periodically to eliminate the bad water which accumulates at the bottom.

❐ Other Appliances

Your refrigerator can move and topple and spill. If it's on wheels, fix them so it won't roll.

Built-in appliances can move out of their enclosures if not secured. It's best to secure them so that they don't. Your dishwasher should be attached to the counter.

All gas appliances should have flexible hoses.

Also, make sure all the pipes and lines in your home are adequately supported. Use plumber's tape. Do not use an iron strap to hold a copper pipe.

If you're not sure how best to secure your appliances, ask the folks at your local building supply store.

❐ Furniture

All furniture which can fall over (on you, a child, a pet) should be bolted to the floor or to wall studs through the back. Metal "L" brackets can also be used. *(See Figure 4.2, below.)* Do the same with bookcases and bunk beds. Modular bookcases and those made from boards and cinder blocks will cause special problems. It may just be best to get rid of these.

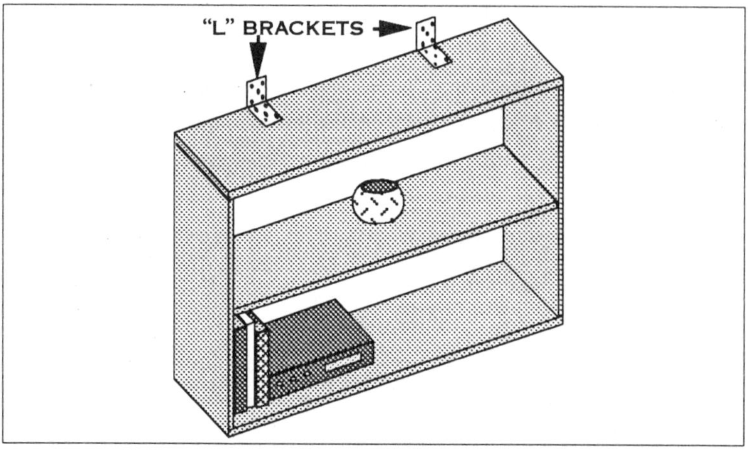

FIGURE 4.2. SHELVING REINFORCED WITH METAL "L" BRACKETS

What a bother! Not only does this take time and money, but may also affect resale value and limits moving furniture around when you feel like it. Is it worth it? It's your decision. Remember that the more things that block your way or block doors, the more difficult and potentially dangerous your life will be during and after a quake.

❒ <u>The Bedroom</u>

First, don't forget that cardboard box under your bed (see ☛ <u>**Mini Kits**</u>, page 59). Remember that you spend one-third of your life in bed. You should have a carpet under your bed. If not, put non-skid coasters under each leg. If the bed is on wheels, take them off or lock them.

The "glass" on the nightstand should be plastic. There should be no mirrors or wall hangings with glass near the bed or hanging light fixtures over it. It's also best to have the bed next to an interior wall and away from any windows.

❒ <u>The Bathroom</u>

Because you're often barefooted here, this room should have <u>no glass containers</u> in it. The medicine cabinet should have a *strong* latch.

If you have an older glass tub enclosure or glass shower door, you should, if financially feasible, install instead the newer types which are of tempered safety glass panels that break into small, relatively harmless pieces. This is a good idea even without the fear of earthquake damage.

❒ <u>Closets, Cabinets and Other Shelves</u>

The key with these is to keep everything in and on them from sliding out and off them. When they aren't secured, an earthquake can produce quite a mess in addition to any expensive structural damage your home may suffer. Imagine dishes, glasses and knick-knacks broken into thousands of dangerous pieces. Mix in with this all the liquid and dry chemicals you store in your kitchen, bathroom, etc., and you have the potential for toxic fumes and a goo that will make your life nightmarish for a long while. The search for your car keys could last several hours (unless you have a spare set in the earthquake kit), hours you should spend in other activities.

First, rearrange all items so the lightest are on top

shelves and the heaviest are on the bottom. Don't store light and heavy things in the same area, as the heavier ones may destroy the lighter ones during shaking.

Chemicals (cleaning fluids, paints, sprays, etc.) should never be in glass containers and should be packed in boxes (on the floor or on bottom shelves) that won't tip or break open. Keep any room that has a pilot light free of combustibles. In the garage, pack all chemicals and flammables in strong wooden boxes and secure the boxes so they won't move.

Closet and cabinet doors must have tight latches. There are a wide variety of these available at all hardware stores and home improvement centers, as well as at marine and RV supply stores. Look for latches which are labelled as childproof (with small children, it's good to have these, anyway), as "passive latches" or as "positive-close" decorative ones. Magnetic catches probably won't hold during a strong earthquake.

For open display shelves without doors and for bookcases, you'll need to install some sort of barrier to stop objects from sliding out. We go to a lot of antique shows where displays of old bottles, steins and china have strings, bars, dowels or some type of guardrail to keep these collectibles from falling on the floor.

It's also possible to find padded shelf liner which inhibits movement. Also, consider buying silicon and other adhesive tapes, special putties or heavy-duty Velcro® to secure objects. Use the same materials to hold down stereos, computers, lamps, vases, and the like. Better for a computer, although more expensive, are special lock fasteners which keep the machine very safe. Best to check with stereo and computer stores for their suggestions. Keep stereos as low as possible.

❒ **Walls and Ceilings**

Pictures and plaques should be padded on the back.

Hang them with threaded hooks screwed into wall studs. More sophisticated devices can be bought at frame shops.

All ceiling lights and fans and hanging plants should be tightly secured to ceiling joists. Hanging plants should only be in light plastic containers.

❒ <u>Windows and Mirrors</u>

Unless the glass is tempered (to break harmlessly) or is plexiglass, windows and mirrors can be potentially very dangerous items. The remedy is to install a clear safety film. This is a tricky job to do by yourself and an expensive one to have a professional put on.

❒ <u>Miscellaneous Items</u>

Firewood and lumber in the garage should be kept in large, secured boxes on the floor.

Pack anything breakable you don't display or use often as if it were to be safely moved—that is, wrap it in lots of foam or paper in sturdy boxes.

Antennas are less likely to fall if they're strapped to the side of the building rather than just sitting on the roof.

Consider trimming or (gasp) taking down old, very tall, shallow-rooted trees (pines, spruces and cedars) which can topple on houses, cars, garages, and people.

Finally: thick carpeting will cushion and save many things which do fall and it will slow the movement of furniture.

Are you having fun yet? Hardly. Have *we* actually done all these things in the house we rent? No. Very few of us will go the complete route, but choosing a few projects over a period of time could make your life—and the lives of everyone else in your house—much safer in the future.

2) Structural Reinforcements

In this section, we'll deal just with living units. What happens to your home's structure during an earthquake is very much dependent on a variety of factors such as the soil it sits on, the slope of the lot, the foundation, construction materials and design, and how well it was built.

Because of its intensity, the 1989 Loma Prieta quake was an important test of the construction techniques which are mentioned in this section. The news is good: houses which had been built to California State earthquake standards (enacted in 1934) withstood the tremor with much less interior and structural damage than did homes older than 1934. However, unreinforced masonry walls and chimneys in both newer and older houses did not do well.Newer homes built on steep slopes or poorly constructed ones suffered more damage than those built well and on flat ground.

Houses are primarily designed to support the weight that is bearing down on the foundation; that is, vertical weight. Unfortunately, earthquakes attack both vertically and from the side; that is, laterally and horizontally. As your house is bounced up and down and pushed and pulled, the weakest links (in terms of material and construction) will be the first to give. It is not unusual for the frame to be shaken off the foundation in strong earthquakes.

To withstand this intense shaking, the various major components of a house must be tied together very well so that it is, for all practical purposes, like a single unit. That is, the roof, walls, porch, and foundation must all be tightly connected. In addition, all walls must be reinforced. This bracing must be done in a way that makes the house more rigid, but at the same time, allows the frame to move enough so it won't shatter.

It is not the purpose of this section to give step-by-step instructions on how to carry out this work. If you would consider doing this work yourself, we strongly recommend references #1, #5 and #6 listed in the ADDITIONAL RESOURCES section, pages 117 and 118.

Most homes in Washington are of a wood frame design. Wood is an ideal material for construction and reinforcement—it is both flexible and lightweight. It is also the best type of construction to resist the ravages of earthquakes.

Most older Washington houses, however, were not reinforced adequately during construction to deal with future quakes. Washington building codes now require diagonal bracing of walls to deal with intense lateral forces—specifically, the very intense winds which are common on the coast and as can also be experienced inland. *(See Figure 4.3, below.)* In place of traditional diagonal bracing, King County builders must use a shorter and more expensive "shear clip," which allows walls with many large windows to be adequately braced.

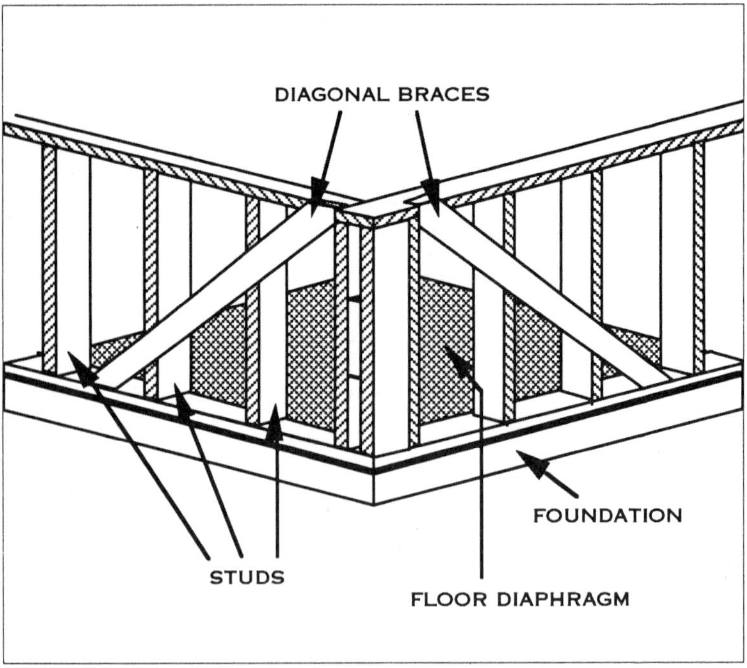

FIGURE 4.3. DIAGONAL BRACING

Although some engineers believe diagonal bracing does more harm than good, most believe that for structures on safer ground and with smaller quakes, it can be adequate reinforcement. But, ideally, it should be used in conjunction with other lateral strengthening techniques.

For example, wall studs should be connected with plywood sheathing, either on the outside or on the inside of the walls. Clapboard siding on the outside of the structure and decorative plywood paneling on the inside can also give extra strength.

Many homes have walls made of short studs between the frame and the foundation. The common term for these is "cripple walls." They create what are popularly known as crawl spaces. Unfortunately, they are quite often poorly braced and will likely collapse during an intense tremor, creating great instability for the frame. Like the other walls of the house, they should be reinforced with diagonal bracing and plywood sheathing; since these walls are often exposed, they can be reinforced without much or any disruption, but this can still be a hard and dirty job.

Since 1985, the State Building Code has required that walls be anchored to foundations. This is probably the most important thing that can be done to protect houses from the shaking of earthquakes.

There are two common ways to accomplish this reinforcement during construction. With one, the foundation is poured with "sill bolts" embedded in it. These extend higher than the bottom sill plates. When bolted, the frame is then tied to the foundation. This can also be accomplished after a house is built, but it's a royal hassle. *(See Figure 4.4, next page.)*

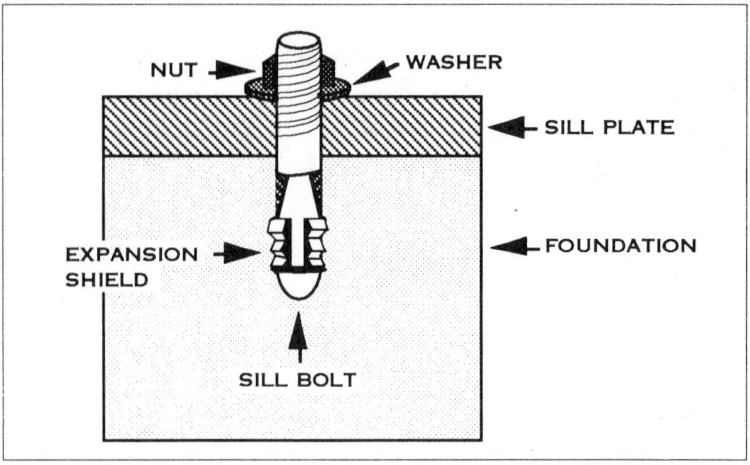

FIGURE 4.4. SILL BOLT

The second approach is for homes with solid concrete foundations to have the frame tied to the foundation with "steel seismic straps." These are metal strips several inches wide and a couple of feet long with holes for nails and bolts that go into both stud and foundation. *(See Figure 4.5, below.)*

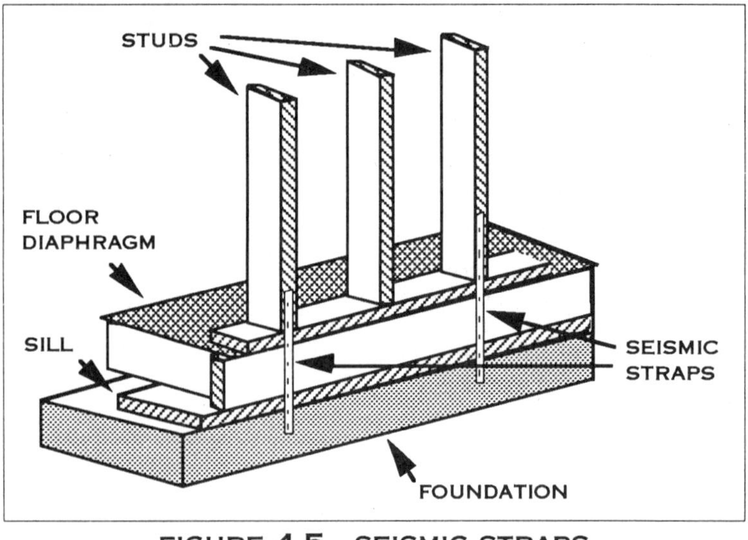

FIGURE 4.5. SEISMIC STRAPS

To create a really tight structure, all connections between studs and sills, between all roof components and between roof and frame should be tied with steel anchoring, nailed plywood sheathing and framing devices. Some common metal connectors are "T" and "L" straps. *(See Figure 4.6, below.)* A wide variety of other connectors is now available which allows almost every possible tie to be made.

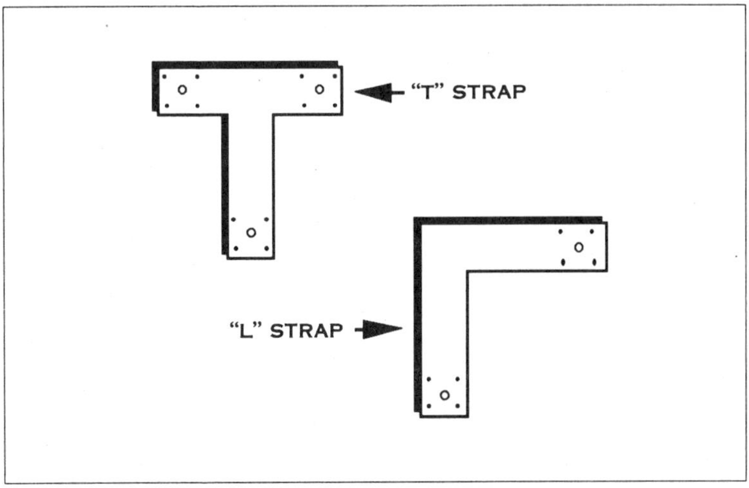

FIGURE 4.6. "T" AND "L" STRAPS

If you have a two-story home, you should also consider tying the two stories together so that the top story won't slide off the bottom one. Some houses are reinforced by individual studs which run from one floor to the next. In a worst case scenario with an unreinforced house, the floors could separate from each other and the frame could slide off the foundation, creating a disastrous situation where the lower walls will give and the upper story and roof will collapse.

(If you want to go state-of-the-art, you might consider a new system that sets houses on a series of springs and shock absorbers. Although much of the earthquake shock is absorbed with these units, financial shock is also a consideration—add at least $15,000 in construction costs!)

Any sort of addition to the basic house—such as garages, porches, chimneys, and fireplaces—should also be reinforced and tied adequately to the rest of the house. Most chimneys are unreinforced and, therefore, present a problem. Lightweight metal fireplaces and chimneys, if anchored properly, are much safer. However, whether it's rock, brick or metal, don't forget that bucket of sand next to your fireplace or wood stove in the event a quake hits while you're toasting your toes.

In most cases, the worst type of building material and design in earthquake country is unreinforced rock or brick. In areas of the world where these houses are the standard, earthquakes will level them and death and injury is very high. In some parts of California, laws have required "retrofitting" (renovating to strengthen the structure) brick commercial buildings or demolition of such structures. Although it's expensive and sometimes unsightly, every effort must be made to reinforce masonry houses.

Stone and brick veneer may look nice on your wood frame house, but it can easily separate, crack and fall, even though the frame may stay intact. It's particularly dangerous if this masonry veneer extends high up the side of your house. Steps can be taken during construction to reinforce this type of facing, but retrofitting can be difficult.

There are also special problems with houses that have a living space over a garage, are on stilts or are split level, as they are inherently less stable. These types of structures have consistently fared poorly in larger California quakes.

Steep slopes that might move a lot are obviously not good choices for a homesite. In addition, some soils, particularly artificial fill and some river deposits which are underlain with water, can actually liquify even during moderate quakes of short duration. We'll discuss soil problems in more detail in Chapters 5 and 6.

Mobile homes, as might be imagined, are easily thrown

off their jack supports and can then be totally demolished in quakes. If at all possible, have these houses secured with a seismic bracing system. Such reinforcement also protects these structures in high winds. For more information, contact a mobile home dealer, your mobile homeowners' association or the Yellow Pages.

No matter whether the construction modifications outlined here are undertaken during construction or as retrofit work, value will be added to your dwelling both in terms of resale and in terms of protection of life and property. If your home is in a high-risk category, insurance companies may refuse to cover it. By earthquake-protecting it, you have a greater chance of coverage.

A good time to retrofit may be when you're planning remodeling or additions or any sort of improvements which require that walls be modified in some way. Of course, all work should be done with a permit issued by your local government building department whether you do the work yourself or hire others. If you're looking for a professional, check the Yellow Pages under "Remodelers," "Contractors" or "Structural Engineers" and make sure they're registered with the State of Washington. In some cases, hiring a geologist or engineer who is knowledgeable about soils might be a good investment because the reinforcement specifics will be determined, to some extent, by the anticipated shaking during a quake.

It may be the case that your house is already partially earthquake-reinforced. Obtain construction drawings for your home to see what has been done to determine what, if anything, still needs to be done. If it's possible to crawl under your home and see where the frame meets foundation, check whether bolts are visible every four to six feet on the sill plate. If so, the frame is anchored.

Having the work done during construction adds several percent to the final cost of the house. However, retrofitting can be much more expensive—running to many thousands of dollars if you don't do the work yourself. Because of this, you may decide to do little or nothing to your home. Most wood frame buildings of one or two stories do fairly well in even somewhat strong quakes and if a structural engineer or home inspector declares your home fairly safe (taking soil conditions into account), you may decide to just concentrate on the earthquake kit, insurance and drills.

Lastly, there is the "renter's dilemma." If you're like us and don't have title to or are not buying your home, there's not a lot you can do except to suggest to the owner that the necessary modifications be made. If you live in an apartment complex, you might call a tenants' meeting and discuss what needs to be done before meeting with the owner.

A note to renters: In some areas, if your dwelling is rendered uninhabitable by a quake, the owner must refund the remaining, pro-rated portion of your rent or provide comparable alternative housing.

3) Earthquake Drills

When you have all the materials necessary for survival in place, you will feel better about your chances for making it through a bad quake with minimal trauma. However, there's even more you can do to reduce stress and reduce the likelihood of personal injury and damage to your possessions.

It's common sense that you are more likely to remain calm and react sensibly during and after a disaster if you have thought about and acted out your responses in advance. Although it's impossible to know what a quake will feel like and what your thinking will be when it actually happens, there's no doubt that drills make a very big positive difference. Just as in fire drills in your own home (you *do* have fire drills, don't you?!), <u>everyone</u> in the home should take part not only in the actions, but also in the planning.

First realize that you won't be able to move very far safely once you're in a quake. Walk through each room of the house or apartment and outside, too, and discuss what would be the most as well as the least safe places to go. Decide which exits would be safe and which would not.

Places to ***go to*** include:

- ✔ A hallway *without* furniture or wall hangings with glass;
- ✔ Under a *sturdy* desk or other furniture;
- ✔ A corner of an inside wall (again, no furniture or wall hangings with glass);
- ✔ An inside reinforced doorway where you can stand or sit holding onto the doorframe.

Some places to ***get away from***:

- ✘ The garage and kitchen (lots of stuff that can fall and fly around);
- ✘ Any glass (windows, mirrors, etc.);
- ✘ Heavy hanging objects;
- ✘ Tall, heavy furniture which could fall on you;
- ✘ Unreinforced masonry structures such as fireplaces, chimneys and outside brick-faced walls.
- ➤ If inside, don't run outside.
- ➤ If outside, move away from trees, utility lines, etc.

The first drill should be at a time everyone knows about. Then do surprise drills. Give one person the responsibility to call out "Earthquake!" anytime during an agreed-upon day.

In addition, everyone should know what to do after the quake, where each fire extinguisher is and how to use it. Practice first aid and CPR on each other and know where the nearest hospital, fire station and police station are. Also, read and discuss the sections at the end of this handbook entitled WHAT TO DO DURING A QUAKE AND AFTERSHOCKS (pages A1 to A7) and WHAT TO DO AFTER A QUAKE (pages B1 to B16) and incorporate these ideas into your drills.

Each person should know how to turn off electricity, gas and water, but one responsible person should be designated to take charge of these duties after a quake. Know how to use water shut-off valves under sinks and toilets. Don't forget to check to see if the main water cut-off requires a special tool. If so, keep that tool attached to the cut-off. Everyone should be familiar with the circuit breaker or fuse box—where it is and how to turn it off. *(See illustration on page B6.)*

Gas valves on gas meters are more complex and very important. Leaking gas is a potential fire hazard, BUT ONCE THE GAS IS TURNED OFF, ONLY A UTILITY COMPANY EMPLOYEE SHOULD TURN IT BACK ON. BECAUSE OF POSSIBLE DAMAGED LINES AND LEAKS, ONLY THE GAS COMPANY SHOULD RELIGHT PILOT LIGHTS. If the disaster is widespread, it could be weeks before the required experienced folks show up.

Check soon to see if the main shut-off valve is accessible. The general rule is: if the handle is parallel to the gas line, the valve is open; if perpendicular, it's closed. *(See Figures 4.7 and 4.8, next page.)* CLOSE THE GAS LINE ONLY IF A DANGEROUS SITUATION EXISTS.

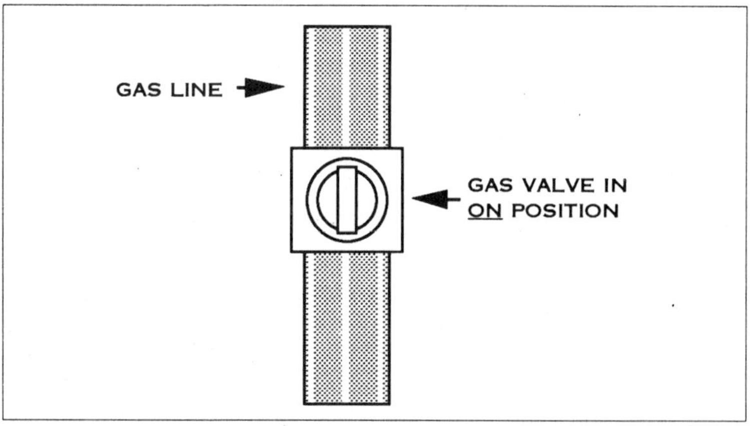

FIGURE 4.7. HOW THE MAIN GAS VALVE SHOULD LOOK NOW: THE OPEN POSITION.

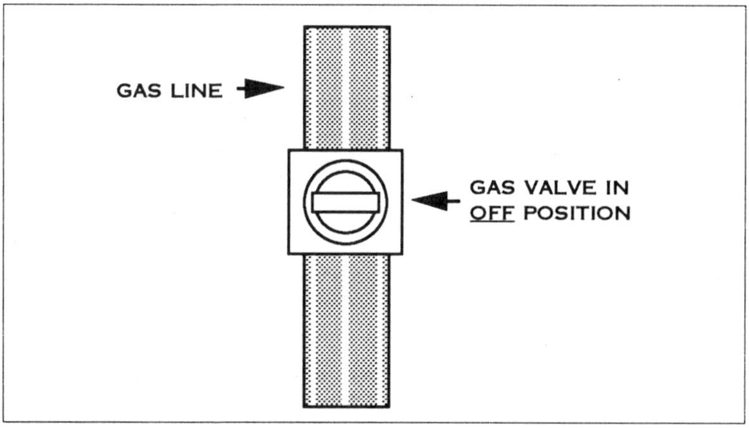

FIGURE 4.8. THE MAIN GAS VALVE: THE CLOSED POSITION

You might consider having an automatic shut-off valve installed on your gas meter. The advantages are that it works when you're not at home and the risk of fire is lessened. In a moderate 1992 Tokyo quake, all valves shut off automatically because, by law, they have been installed in every Japanese home and building.

However, there are problems, too. One is the expense—$200 or more. Another is the possibility that they will turn off even when there is no damage to your gas line, but you won't know if there *is* any damage until a qualified utility worker arrives and you will be without gas in the meantime.

Washington Natural Gas customers will be happy to know that the company has upgraded almost all of its gas lines so that they are flexible and much less likely to break during a quake.

There are a few more things to do in conjunction with drills:

· Check to make sure that all family members' immunization shots and records are up to date. The spread of disease often increases dramatically after disasters. Medical personnel will be overworked and medical facilities may be damaged.

Everyone should realize and accept the possibility that if the family is separated, it could be many hours or even days before everyone can safely return home. Since not everyone may be at home during the disaster, you should decide on one meeting place after the quake. You should also decide on an alternate meeting place in case authorities require you to evacuate your neighborhood. Get in touch with your town's emergency management bureau to find out where evacuation areas are planned near your home.

Children home alone should have a neighborhood house to go to. If a young child is at school, they should know to stay there until a friend or relative comes to get them and that it might be a very long time before they show up.

Since anyone can get panicky or depressed that drills are even necessary, stress the fact that drills are the best way to ensure that everyone will end up okay. Emphasize that, on average, only a few people die from quakes each year in the U.S., while tens of thousands die in auto accidents. If you do your drill at least several times, children in particular will be less likely to panic when a quake actually hits.

Pets often run away during a catastrophe and you should realize that they may not come back. Also, if you are ordered to leave your home and go to an evacuation area, you will not be allowed to bring your pets to the temporary shelter, so you should put them in a room that is well-stocked with food and water and to which you'll be able to return within a few days.

Contact your neighbors to set up a block or neighborhood meeting where you can coordinate supplies, education and activities such as periodic drills. Learn who has medical and other pertinent skills and think about how people should take responsibility in case neighbors aren't home.

Also, bring up the issue of disasters at your workplace. If there isn't one, strongly suggest that an emergency plan be devised and that earthquake drills be conducted.

Finally, if you live on the coast or near any large body of water, your drills should include what to do in case of a tsunami or flooding (see page B16 in the WHAT TO DO AFTER A QUAKE section).

4) THE ELDERLY AND THE HANDICAPPED

These two groups share in common the problem of often not being able to respond to danger as quickly as other people. For this reason, extra precautions should be taken. If possible, everything large and moveable should be secured in the house or apartment, particularly medical equipment and dangerous items. It may be more difficult for the handicapped and elderly to move out of the way of falling objects and more room may be needed to safely evacuate and move around the structure once the quake is over.

It would also be a very good idea to put in each room an automatic power failure emergency light which turns on when the electricity is cut off and, if hearing is a problem, smoke detectors which emit a flashing light as well as a sound.

Just as important as physical security is the development of a buddy system for those living alone. Friends, relatives or volunteers should be paired with those who might have extra difficulties. As soon as a quake is over, the buddies should check up on each other.

The elderly or handicapped should always carry on their person a whistle, a list of necessary medications with copies of prescriptions, instructions on how to deal with their medical problems, and phone numbers and addresses of physicians, clinics and relatives. The buddy should also have this information.

If you live in a building designed to house the elderly or handicapped, ask the management about earthquake safety and evacuation plans. If there are none, stress that they must be developed.

5) Earthquake Insurance

It's not the goal of this handbook to promote insurance companies or any individual agents. However, since a major quake in Washington is not really a question of if but when, you may decide that earthquake insurance is right for you.

It's a rather amazing fact that only about one-fourth of all California homeowners have earthquake insurance. Some California geologists and insurance agents don't even have coverage! And only 7% of Washingtonians carry an earthquake endorsement (extension) on their homeowner's insurance. An even smaller percentage of renters have this coverage.

First, let's discuss the situation you'd be in if a big quake rocked your area and you weren't covered. If you're in a good house on good soil some distance from the epicenter, your home may not sustain much damage, perhaps less than the standard 5% or 10% deductible which comes with all endorsements. However, if your house and possessions are heavily damaged, you may not have the resources to rebuild. The federal government may provide some aid to tide you over—a grant to pay for lodging for up to one and a half years if your home is not liveable and a grant to replace up to $10,000 worth of your possessions. In addition, you may also be able to get a loan (not a grant) of up to $100,000 to repair your home or to rebuild. You could also receive tax refunds or tax relief.

The worst case, no-earthquake-insurance scenario would be if you were buying your house and it suffered so much damage that it had to be demolished. If you then rebuilt with a government loan, you'd be paying two mortgages, but would only be buying one house!

An earthquake endorsement for a wood frame house is fairly inexpensive in Washington, only about 75 cents per $1,000 coverage; however, the premium for a stone or brick house is about $2.25 per $1,000. If you want total coverage, insure at the replacement cost, not the market value, which may be different in cost. This holds for insuring personal property, too.

An earthquake endorsement can only be added to your present policy by your home or renter's insurance agent. Although not required to offer this coverage, almost all companies do for competitive reasons. Also, most companies won't insure your home for earthquake damage if it's built on piers or stilts, if it sits on landfill or if there's problems with the foundation. If you're denied this coverage, you might ask what, if any, modifications of the structure are necessary to make it eligible.

Your regular homeowner's (fire) policy covers all earthquake-related fire and smoke damage. But only the earthquake endorsement covers quake-related structural damage and water damage due to broken water pipes, tsunamis and earthquake-caused flooding. It will also pay reasonable additional living costs if your home is not liveable after a quake. Your car is insured only if you have the optional comprehensive coverage which insures for damage due to falling objects and fire.

In the notebook which you keep with your earthquake supplies, make a list, room by room, of your possessions and their replacement costs. Review this inventory each year. If you have the time, money and/or inclination, you might also videotape or photograph each room. Keep extra copies of photos and the inventory in your safe deposit box.

CHAPTER 5

DID THE EARTH MOVE FOR YOU, TOO?

SOILS, BUILDINGS AND COMMUNITY SAFETY

Thanks to ancient volcanic and glacial activity, spectacular mile-wide rivers, epic flooding, vast primeval forests, and relentless ocean activity, western Washington's scenery has an almost mythic intensity. Because of these influences, the soil is moist and rich. But for the most part—unfortunately—this soil is very likely to fail during large earthquakes.

For a variety of reasons—for example, ease of construction, bigger profits or just ignorance of the obvious—erection of homes and buildings has taken place on flat flood plain and fill soils or on dangerously steep "view" slopes. And, unlike the cultures which preceded us, we have increased our numbers without control, covered some of the best ground with concrete and asphalt, dammed and bridged the rivers, gouged out the hills and mountains, drained the wetlands, felled the forests, diverted and covered the creeks, and filled storage tanks to overflowing with vast quantities of the most dangerous substances known. If there truly is a female nature consciousness and if she could speak in human terms about the next great cataclysm, her words might well be, "Okay, boys, it's pay-back time!"

What a fine mess we've got ourselves into and now we have to face up to and deal with our mistakes. Firstly, we have to regulate construction and destruction much more and in different ways than we have in the past and, secondly, we must modify our existing structures—houses, office buildings, schools, hospitals, roads, and bridges—so they are less likely to be damaged in Washington's first great quake in recorded history or even in lesser tremors. Fortunately, much work is already underway to accomplish this: the ground is being carefully studied and building codes have been changed to fit in with this new-found reality.

SOIL AND SLOPE

Your safety in or on a structure during a large earthquake is primarily determined by two factors: the adequacy of that structure and the type and slope of the soil and rock below it. In this section, we'll discuss ground conditions.

The intense shaking of an earthquake can produce ground failure such that the things sitting on the ground—houses, bridges, buildings—may collapse and it can also produce ground movement which slides around and breaks up the structures. In addition to the type of soil underneath the structure, the amount of shaking is also determined by the magnitude of the earthquake and the distance from the epicenter—in general, the larger the quake and the closer you are to it, the greater the shaking.

In a very general way, we can also rank ground type from most dangerous to least dangerous as follows:

Increasing ground vibration during an earthquake ↑	
	Poorly packed artificial fill (most dangerous)
	Marshy lands and mud
	Loose sand and gravel (from ancient stream and river beds and flood plains)
	Sedimentary and other rocks and certain types of rocky soil (least dangerous)

There are, of course, many types of ground which fall between these classifications and which may have characteristics of two or more of these. Also, in general, for any type of ground the greater the slope, the greater the risk. Slopes which have grades greater than 25% can be very landslide prone, while slopes less than 15% rarely are. However, houses on very steep slopes can be engineered to be quite safe even though they may not look it.

A significant number of structures in Washington have been constructed on the more dangerous ground. In addition, much of this land is near water—creeks, rivers and oceans—and has yet another dangerous feature: a wet, sandy soil layer which is often near the surface. If shaken intensely for at least ten to twenty seconds, the water is squeezed from this layer and permeates the soil above, causing a quicksand-like mess known as "liquefaction."

When ground has liquefacted, it then moves in such a way as to crack foundations and roads, break underground pipes, and move and tilt houses and buildings. It was this condition that led to the sinking and breaking up of houses in the Marina district of San Francisco in 1989. The shaking lasted only a few seconds, but the ground failure resulted in three-story wood frame and stucco houses sinking and collapsing, with only the top story standing when the quake stopped! Ironically, these houses had been built on artificial fill originally developed to hold buidings for a fair which celebrated the rebuilding of San Francisco after the 1906 earthquake and fire.

Coastal areas of Alaska suffered horrible liquefaction-caused destruction during the great 1964 subduction quake. Inland, some hillside homes in Anchorage traveled <u>up to a quarter of a mile</u>, probably as a result of land movement produced by liquefaction!

A large proportion of Washington's population lives, works and drives on such potentially liquid soil.

Ideally, the best areas to build on are shallow soil layers which sit on solid rock. Of course, such conditions are usually on hill or mountain slopes, which may not be good. In such areas, builders will often dig out a building or road site and use the debris as a fill on which to build part of the structure. This fill is sometimes unstable and landsliding can occur in the weakened area above the cut. In addition, plants and trees which help hold the soil are often taken out in development.

Even without these practices, rock base slopes can be landslide prone because the rock has broken and loosened naturally over time. Major quakes in Washington (greater than a Richter scale magnitude 6.5) will result in many rock slides which will close important roadways. Roads through the Cascades and Olympics are at great risk because of steep slopes and slide-prone soils.

A few more notes about soil dynamics and risk:

1. The stability of sea cliff rock and soil during quakes is quite poor.

2. Structures should never be built on known faults which have the potential to be active, as even the best-built structures will likely be destroyed or damaged. Some geologic maps show the locations of these faults.

3. Soil at the base of hills and mountains is often very dangerous. Earthquake shock waves travel through this softer soil, but then bounce back when they hit bedrock, magnifying the intensity of the shaking.

BUILDINGS AND BUILDING CODES

People who live in earthquake country appreciate the saying, "Earthquakes don't kill people, buildings kill people." But thanks to a tremendous amount of experience and technology which has resulted from quakes in Japan, California and elsewhere, engineers worldwide now have the know-how to build much safer structures. Although it is impossible to construct buildings which sustain absolutely no damage in a very strong quake, structures which sustain no dangerous or heavy damage are quite feasible.

The effects of earthquakes on buildings are very complex. Architects must work with structural engineers, seismologists, soils engineers, and geologists or with specialized earthquake engineers who are knowledgeable in all these areas.

Because this is a handbook for individuals and families, a detailed description of construction techniques to make buildings earthquake-resistant is beyond the scope of this work. However, some facts and figures will help us get a better grasp of the situation.

We've mentioned how important it is to understand soil conditions for making any structure safe. Larger, wider structures have an extra complication: their foundations often sprawl across a variety of geologic conditions and soil types. In addition, many buildings are multi-story and are constructed of very heavy materials. Therefore, before any work is done, samples of rock and soil must be extracted by means of site drilling.

Buildings sitting on bedrock need to be constructed quite differently from those built on soft soils. The thickness of the layers of soil and how loose or compacted they are also influence how the building vibrates and, therefore, influence earthquake design.

For many types of soil, the thicker the soil down to bedrock, the less the shaking. Very thick soil layers can absorb much energy below the surface.

Soil can be engineered so as to be safer. Soft soils, as we have mentioned, can liquefy because of subsurface saturation. However, drainage systems can be installed to deal with this problem. In addition, there are machines which can compact loose soil. These techniques can also help cut down on landsliding.

Artificial fill is another problem which must sometimes be dealt with. Many types of fill, such as sawdust and garbage dumps with lots of organic matter, are inherently unstable and should not be considered as building sites. However, if the fill is composed of denser materials—for example, concrete and bricks—it is often better than some naturally deposited soils.

Buildings of different heights respond to shaking in varying, but distinctive, ways. Buildings, like houses, must be adequately tied together. That is, the foundation must be firmly secured to the frame which, in turn, must be tightly connected to the roof. The quality of execution of detail and the skill employed in putting the building together are other essential factors.

If masonry is used, it must be carefully reinforced and connected, otherwise it is potentially the most dangerous of building materials. Reinforced concrete can be used to good advantage in making a building earthquake safe. However, steel frame structures are the safest. Like wood, steel easily flexes and it is, of course, stronger than wood. Steel also has the advantage of being much lighter than concrete.

The possible frequency and intensity of shock waves and the effects of aftershocks are other factors architects and engineers must take into account. A strong quake can shake a structure up to twenty times, while a smaller quake will shake a structure only once or twice. The Loma Prieta quake destroyed only a few buildings, but experts estimate that if it had lasted only several seconds longer, a much larger number would have collapsed. A future long duration Cascadia subduction quake will certainly put all our structures to the ultimate test.

To do all the things necessary to make our buildings earthquake resistant would be very costly, running into many billions of dollars. Although modifying structures under construction only adds several percent to the bill, retrofitting can be much more expensive. However, a very large quake could be much more costly, not only financially, but also in terms of injuries and lives lost. Monetary losses in the 1989 California quake topped $10 billion and this in an area that has some of the toughest codes in the country. To stress the point once more: Washington could be hit with a much larger, longer-lasting tremor and the state is at greater risk because a large percentage of buildings have not been designed with very intense, long-duration earthquake shaking in mind. Economic losses as a result of a very large quake centered in Puget Sound could easily surpass $10 billion and estimates for a great subduction quake are in the many tens of billions for the Pacific Northwest.

To begin reducing the risk, there are various things that can be done. One is to strengthen existing buildings which are considered to be the most important facilities to be operating after a quake. Fire stations, hospitals and power facilities obviously rank at the top of such a list.

Building codes should always be re-evaluated to see if new construction should meet tougher standards. With each careful study of the effects of quakes, scientists and engineers expand their knowledge.

Although a building code describes minimum requirements, it's best to build for the maximum potential quake and other hazards. For example, even before there were statewide earthquake provisions, many municipalities and engineers did design some buildings with high winds and earthquakes in mind.

In addition to these building code standards, the other ingredients for safety are money to fund the work and qualified professionals to carry it out.

If we're lucky and have, say, fifty years before a major quake, simple attrition of older buildings and the construction of new, earthquake-resistant ones will make the lives of Washingtonians much safer. But, at the present, there are lots of aging major structures which have not been "weeded out" over the years by more frequent earthquakes, as happens in other parts of the world.

Nevertheless, there are some hopeful developments:

➤ In 1994, all of coastal and southwest Washington, the Vancouver metropolitan area and certain adjacent areas were reclassified to the same seismic risk zone as the Puget Sound area (Seismic Zone 3). As a consequence, building codes will be upgraded and new buildings and houses will be earthquake safer.

➤ Since 1991, codes have incorporated tougher standards for the foundations of critical facilities such as schools, hospitals and fire stations.

➤ Local jurisdictions can make state codes more stringent, but cannot weaken them.

➤ The most populous and fastest growing counties and cities must include geologic hazards in their comprehensive plans.

➤ Washington state public buildings are retrofitted as the need arises and as funding allows.

CHAPTER 6

ESSENTIAL SERVICES: THE ROLE OF GOVERNMENTS AND OTHER ORGANIZATIONS

Much of the responsibility of planning for and then dealing with the aftermath of earthquakes falls on the shoulders of federal, state and local governments. Indeed, in surveys in California, the overwhelming majority of respondents felt that earthquake preparation is one of the primary duties of government.

In fact, state, city and county governments must, because of their role in overall social planning, development and safety, take this leadership. Before a quake, they must deal with reducing risk; immediately after, they are expected to provide emergency assistance; somewhat later, it is their duty to restore critical services and facilities; and for the months and years following, they help direct the rebuilding of their communities.

In addition, an effective response requires that the various governments and all the agencies within them coordinate their plans and activities. The number of necessary interactions between these entities becomes, if not impossible, at least mind-boggling. An encouraging sign is that the will exists to surmount these problems. For example, an earthquake conference in Seattle in 1990 drew over 1,000 participants!

The 1989 California Bay Area quake was one test of the capacity of an intergovernmental network that had been developed over many years. It was generally felt that the system performed adequately, but that a lot more effort would be required to get it working at an ideal level.

Although Washington's earthquake preparedness program has been developed over a period of years, government officials and workers are always seeking information from sources outside the state. For example, the Federal Emergency Management Agency (FEMA) has a variety of organizational guidebooks for local government. In Northern California, the Bay Area Regional Earthquake Preparedness Project (BAREPP), begun in 1984 and operating out of the Governor's Office of Emergency Services, is able to share its knowledge—which is the result of extensive work in a regional twelve-county area—with public as well as private and volunteer organizations. (See listing #10 in the ADDITIONAL RESOURCES section, page 119.)

However, Washington's governments and agencies, unlike those in California, face an initial problem of trying to convince much of the population that preparation is even necessary. Skepticism that Mt. St. Helens could dangerously erupt was widespread in the Pacific Northwest until the volcano actually exploded. Geologists were impressively accurate with their predictions then, but similar predictive success with earthquakes is highly unlikely.

In addition, some defense mechanisms come into play when the reality of big quakes in Washington is finally accepted. Many people will deny that their town, neighborhood, house or workplace will be hit. Another outlook which must be overcome is fatalism: "It was meant to be, there's really nothing we can do about it."

An additional problem is that some government officials might be unwilling to act. A few may simply not be educated about the earthquake danger. Others may see earthquake preparedness as a headache that won't bring them any rewards (or votes). After all, the bearer of bad news is often the one who is blamed for the news, so why not just make it through the term and leave the tough choices and hard work for someone else.

Others will worry that preparation will lead to panic (even though the evidence argues that just the opposite is the case). And yet others will be influenced to keep quiet by those who may lose money (or at least not make more of it) because of land use decisions and the like.

But because the evidence for future great quakes in Washington is so compelling and because public awareness is growing, greater government involvement in earthquake preparedness planning is necessary and inevitable. The remainder of this chapter will outline what is involved with this planning and the degree of organization at present.

BEFORE...

One of the early activities communities must become involved with is risk assessment. This includes identifying which structures and areas are potentially hazardous. First, seismic studies must be conducted on how the ground will move during a quake. In Washington, federal and state investigators have conducted such studies throughout the state. The Pacific Northwest also has a large number of seismograph stations, most operated by the University of Washington.

With this knowledge, officials can then determine which neighborhoods, buildings and bridges are in greatest danger. Next, new building codes, zoning laws and construction standards are devised and mandatory retrofitting programs, like California requires of many older, unsafe buildings, may be developed. Facilities which are essential to normal functioning—such as power stations and lines, water storage and pipes, bridges, roads, sewage plants and systems and communication facilities—must receive special scrutiny and tough standards because all will need to perform adequately for successful disaster relief after a quake.

Because it is best to prepare for the worst possible scenario, all facilities which could add to the danger after a quake—for example, those containing hazardous materials which could leak or explode—must be regulated and modified or shut down.

Emergency response procedures are developed in conjunction with hospitals, fire and police departments, the Red Cross, schools and volunteer organizations such as neighborhood associations. Periodic drills are necessary to test the effectiveness of the plans. Emergency shelter and care areas are designated and relief supplies are stockpiled.

To help with the funding of these processes, Washington state government receives money and other assistance from several federal agencies, including FEMA. Some of these funds are distributed to local governments. With the possibility of state government spending cuts, federal funding is even more crucial for the work that is so essential in the future.

However, a variety of activities are underway in spite of tight funds. The Washington State Department of Transportation is engaged in a substantial bridge and highway retrofit program. Emergency management exercises have been conducted by state and local jurisdictions to prepare and evaluate various agencies' abilities to respond to a major earthquake and coordination between agencies. These include mock rescue efforts and response to the loss of highways, energy sources and the like.

In addition, conferences to help educate those involved in preparedness occur on a regular basis. These involve people as diverse as architects, fire fighters and utility company representatives.

Government is also in the business of education. Public awareness campaigns educate not only about the reality of the danger, but also steps individuals and groups can take to reduce the danger. Neighborhood, school and church groups are essential for the dissemination of this information. Libraries should carry extensive earthquake material, including a selection of very good videos which are presently available.

All American Red Cross chapters distribute a wide variety of information for the general public relating to disaster preparation in general and often earthquake information in particular. These agencies also often provide knowledgeable speakers to give presentations to community groups and businesses. In addition to offering standard first aid and CPR classes, Red Cross chapters may offer disaster services training classes for individuals who would like to assist in mass care work after a disaster. The King County–Seattle branch (listing #7 in the ADDITIONAL RESOURCES section) is active in emergency and educational networks and helps organize interagency conferences and vendor fairs which are open to the public.

The University of Washington Geophysics Program has a public information telephone line (206 543-7010) with recorded updates on recent earthquakes.

Money might also be well-spent in purchasing vans which simulate earthquakes. Popular in California, they can travel to schools, neighborhoods and workplaces and allow people to go through a simulated earthquake. After such an experience, most folks want to learn more.

Citizens are more likely to take the threat seriously when public officials, service organizations and scientists make joint statements about earthquake danger, particularly during April of each year when one week is designated statewide as Earthquake Awareness Week.

School districts and care centers need to be required by state law to hold regular earthquake drills. They should also develop earthquake emergency plans, with specific procedures to carry out the plans. In addition, each school within a district should develop its own plan related to its own specific circumstances and resources.

All teachers, administration and other staff at each school are responsible for some aspect of preparation and have specific duties to perform during and after a quake. It is also a good idea that students be given responsibilities and that both parents and students be encouraged to take part in the development of the school emergency plan.

Experience has shown that carrying out a plan is most successful when each school has several committees which act as response teams; for example, fire safety, evacuation, search and rescue and first aid, and that one person act to coordinate these groups.

Emergency supplies and equipment should be stockpiled and checked and all hazards inside and outside school buildings must be identified and remedied. Guidebooks have been developed for Washington schools which help them develop all these activities.

Government must also take the initiative to require both public and private schools to strengthen schools buildings so that they will not collapse during a quake. Some school districts around the state are actively involved in school retrofitting.

...AND AFTER

In the hours and days after a quake, governments can be somewhat immobilized because of destruction to some of the critical facilities necessary to deliver emergency relief. To a large extent, we could be on our own as individuals, families and small neighborhood groups. Lots of volunteers will spontaneously initiate rescue and clean-up efforts.

Washington has sixty-nine local emergency management jurisdictions in addition to a state agency. Projects are now underway to organize many areas of the state by neighborhood so that groups of volunteers are in place before a quake to deal with education and preparation and afterwards to deal with response activities. Kitsap County, for example, has dozens of organized schools and neighborhoods. And Puget Sound Power and Light has paid for materials and other costs for initial neighborhood training for workers in all the emergency management agencies in their service area.

Governments and businesses are not the only organizations taking these initiatives. A citizen initiated, based and directed group, Mothers for H.E.L.P. (Help Everyone Learn Preparedness) has been in the forefront of education and preparation for several years (see listing #8 in the ADDITIONAL RESOURCES section.)

A prepared public takes some important burdens off larger coordinated relief efforts which aim to deal with critical facilities and infrastructures.

For at least 72 hours after a disaster, the Red Cross will not have information on the health and welfare of people living in the affected area since all efforts will be directed toward feeding and sheltering survivors. Even when the Red Cross is ready to take such inquiries, you will only be allowed to inquire about family members. If you are concerned about a friend, get in touch with their family.

Many local government employees, including teachers and other school staff, will, when possible, be authorized as relief workers. During this time, some workers will be attempting to deal with fires, making repairs on bridges and roads, clearing debris and the like. It is crucial that large-scale relief start as soon as possible because individuals will become exhausted and food and safe water will become scarce.

Temporary shelter in the form of tent cities or in safe public buildings will be created with the help of the Red Cross, National Guard and perhaps other military, depending on the size of the disaster. Information about the general state of affairs will be transmitted primarily through the Emergency Broadcast System, while ham radio operators will dispense specific local conditions so authorities can better plan relief and rescue.

Government's final role is to help bring life back to normal. No less than twenty-five (!) U.S. government agencies will perform various support and rebuilding functions in the weeks and months after a quake. This return to normal will take many years. Even several years after the Loma Prieta quake, much still remained to be done to repair buildings and roadways in San Francisco, Oakland, Santa Cruz and Watsonville.

Local government services must be re-established, transportation and utilities must be restored and hazards must be identified and dealt with. Because all this is very expensive and is virtually impossible for any hard-hit community to finance on its own, the federal government, as we've mentioned, gives grants and underwrites low-interest loans to some individuals and businesses and gives matching funds to rebuild public facilities. Since these grants often fall short of the need, local taxation is often used to make up the difference. To raise some repair money after the Bay Area quake, San Francisco voters passed a bond measure funded by a temporary sales tax increase.

Government and private agencies must also deal with the psychological scars left by a disaster. Damage to homes, injury and death of others leads to tremendous stress which is expressed as depression, nightmares and irritability, particularly in children. Immediately after the quake, mental health services will begin to seek out the traumatized and provide therapy on the spot. In time, those requiring help can receive it at mental health centers.

So far, this chapter has outlined statewide planning. The remainder of this chapter deals with local planning, specifically the situations in the Puget Sound area, on the Washington coast, and along the Columbia River.

COASTAL BAYS, HARBORS AND SHORELINE

Much of the recent evidence which has led us to the conclusion that all of the Pacific Northwest has experienced many great quakes in the past and will suffer more in the future has been observed along the coast.

Proof is found in the buried marshlands all along the coastline. As we mentioned in Chapter 3, subduction quakes often result in the ground dropping several feet or more. Coastal marshes are then inundated with salt water from great waves and flooding which kills plants and trees. And, as mentioned earlier, from highway survey and laser measurement data, we have also learned that the area is rising and tilting, most probably as a result of strain produced from locking of the Cascadia Zone plates.

The coast has some problems peculiar to it which makes the whole length of it quite risky, particularly considering that the source of the next great quake will probably be on or near the Cascadia Zone, shaking the coastline more than other parts of the state.

First, many of its communities are rather isolated. The roads and bridges which connect them with the interior will almost surely be impassable as a result of landsliding or buckling. For example, the Olympic Mountains are formed from the mud and sedimentary rock being scraped off the Juan de Fuca plate as it slides underneath the North American plate. This mountain soil is unstable even under the best of conditions and will surely cover the roadways during and after a big quake. Increasing the landslide potential throughout the area is the very high annual precipitation, steep slopes and the widespread destruction of natural erosion control as a result of logging and development.

Second, many homes, buildings, port facilities, primary service facilities and industrial areas are resting on poor soil. Some towns are built to a large degree on sand saturated with water. Others sit on bay mud and other loose soils. Soil liquefaction on the coast and along bays, then, is a frightening possibility. (Some areas are even now experiencing major beach erosion problems while lots continue to be sold and developed.) Increasing the danger is the likelihood that industrial chemicals will probably pollute the air, water and ground, making environmental and economic recovery difficult and also increasing the chances of ruinous fires like those that devastated Seward and other Alaskan coastal areas after the 1964 quake and tsunami.

Because of the relatively large population and also because of its economic importance, the Grays Harbor (Aberdeen–Hoquiam) area has been studied extensively to anticipate various earthquake effects. Most of the population of this region lives and works on soil that will almost certainly liquify during a strong, long duration quake. Shoreline residential areas like the town of Ocean Shores and waterfront industrial areas will permanently sink five feet or more. The long term effects of economic disruption may be devastating.

Finally, as if things weren't bad enough, tsunami damage is almost a certainty. Since the origin of these very fast-moving waves is located not far offshore, precious little escape time will be possible. However, if only one segment of the Juan de Fuca plate breaks, areas farther away will have more time to evacuate. When a major quake occurs in distant areas, elaborate preparations can be made for a possible tsunami. (Eight hours after the giant Alaska earthquake, three tsunami waves hit the coast north of Grays Harbor, resulting in major damage.) All coastal regions subject to danger will be warned of possible waves and expected times of arrival.

Hitting the coast in as little as ten minutes after the shaking stops, and up to thirty feet high, a tsunami can cause more damage than any other earthquake effect. The waves can bring down weakened structures and anything else in their path and throw the debris against people and buildings far from the shoreline. Extensive flooding of low-lying areas may take place, lasting up to twenty-four hours. Waves can rush up the coastal rivers many miles. Sea cliffs will collapse from the battering. All these effects will be even worse if the waves hit at high tide or in the summer when many towns are crowded with visitors.

So what steps can be taken to lessen the damage? Harbors and bays may be protected from smaller tsunamis if they have narrow openings or breakwaters. Otherwise, they may actually amplify the effects. It is also important to limit development in risky areas. Taking out the dunes, the shoreline trees and shrubs and the other natural barriers which help lessen the impact of the waves should be banned. At present, there are some state imposed limitations on development which are administered by local governments. Another top priority is to strengthen all bridges to the highest possible standards.

Education is, of course, a top priority. Because the coast has been hit in recent times by tsunamis generated in Alaska and other parts of the world and because it is an active seismic area, tsunami awareness is fairly good. For example, many coastal residents have access to detailed survival guide sections in their phone books. Information includes basic family preparedness, tsunami danger and evacuation routes.

Media coverage has been very helpful in getting the word out. For example, several years ago the Aberdeen *Daily World* printed a special earthquake section which could be saved and used for education and preparation.

Another approach has been to put information directly in the hands of residents and visitors. On a recent Fourth of July weekend, tens of thousands of leaflets which described what to do in case of earthquakes and tsunamis were handed out to each vehicle driving onto the Long Beach Peninsula.

As with other parts of the state, public and private agencies in coastal communities and regions are working together to develop plans to deal with earthquake related dangers. Issues include tsunami alert warning systems, evacuation planning, methods to deal with tsunami caused fires, school safety and public education. In Pacific County, representatives from the fire and transportation departments and from schools meet monthly with elected and other officials and emergency responders to coordinate activities.

Equally important are the federally and state sponsored coastal exercises, workshops and conferences which educate and train those who are responsible for emergency preparedness because much planning and coordination remains to be done.

(Check page B16 of the A*PPENDIX* *of this handbook for specific instructions on what actions to take if you're on the coast when an earthquake hits.)*

P.S.—JUST HOW SOUND IS THE SOUND?!

Historic Washington earthquakes have occurred throughout the state, many in sparsely populated areas. The largest recorded was near Lake Chelan in 1872 and has been estimated at about magnitude 7.5. The northeastern region, which includes Spokane, has experienced almost no seismic activity. Most quakes occur in northwestern Washington, under Puget Sound and in the Olympics and Cascades. Recent evidence suggests a one hundred-plus-mile-wide zone of active faulting which extends from the Olympic Peninsula southeasterly down through the Hanford and Tri-Cities area.

The most dramatic and deadly quakes in recent Washington history occurred in Puget Sound in 1949 and 1965. Measuring magnitudes 7.1 and 6.5, respectively, they were both felt over very large areas and caused damage well over one hundred miles from the epicenters.

Because of rapid population growth over the last few decades, most of today's residents did not experience either quake. For this reason and because of the unfortunate but understandable tendency to push potentially negative information from our conscious minds, it's important to remind ourselves of this history. The earth dynamics which led to these large quakes have not ceased and will lead to similar events in the future. Best guess estimates are that the area is due, or even overdue, for a magnitude 6.5.

Each earthquake created extensive building and home damage, resulted in liquefaction and landsliding, and led to several deaths and many injuries. Both were a result of activity in the subducting Juan de Fuca plate many miles below Puget Sound. For this reason, there was little aftershock activity and, therefore, less damage than if similar breaks had occurred closer to the surface. In today's dollars, each quake caused several hundred millions of damage. Comparable shaking today would lead to a much greater loss of life, injury and destruction because of a larger population and more extensive development.

Centered a few miles east-northeast of Olympia, the 20-second April 13, 1949, earthquake hit several minutes before noon. Luckily, schools were on vacation and most workers had not left for lunch, for hundreds of thousands of bricks rained on sidewalks and schoolyards throughout the region. In Chehalis, the majority of the town's chimneys were demolished. Significant damage also occurred in Longview and Seattle on loose fill and sandy soils. Tacoma was also shaken hard, with a giant landslide in the Narrows as the most spectacular result.

The epicenter of the April 29, 1965, Puget Sound quake was just south of Renton. The greatest damage, however, was on Harbor Island in Seattle and in West Seattle because of the very poor soils in those areas. Tacoma and Olympia were again hard hit. As with most quakes, a few seconds more shaking and the destruction and death would have been significantly greater.

At this time, losses for a 7.5 quake below Puget Sound while schools are in session are estimated at over 2,000 dead, with many times that number injured and homeless. An earthquake of this size is larger than historic quakes but, considering the configuration of the subducting plate, is certainly feasible. The plate segment under the Sound "rides higher" than the southern segment under Oregon and the density is also greater. According to scientists, this combination explains why subducting plate quakes are more common and can be larger under Seattle than under Portland, for example.

The Sound area also suffers from unusually dangerous soils and slopes. Most of this problem has to do with an ancient history of glacial and earthquake activity and a legacy of river and bay sediments. Glacial deposits in ancient gouged out basins are often two to three thousand feet thick below Tacoma and Seattle. Most recently, human development has added to the problem by artificially filling rivers, streams and shorelines.

Geologic hazard mapping of the area has advanced

dramatically over the past several decades. The results are sobering. A significant portion of low lying areas have a high likelihood of liquefaction. For example, the valleys of the White, Cedar and Green Rivers are all in this category. In addition, in an effort to expand the land base, much of the Renton, Kent, Auburn and Tukwila areas have been filled with very loosely compacted materials in the last century. This is particularly unstable soil. Much of the rest of the land in the region consists of steeply sloping bluffs which can easily landslide. If this occurred in lakes or reservoirs, flooding could also be a serious problem.

In recognition of this danger, municipalities and counties have been regulating construction and requiring more sophisticated soil analysis reports in these sensitive areas. For example, King County regulations prohibit development or require very well engineered structures on very steep slopes, on wetlands and on soils which will experience considerable liquefaction. Land use planning is also taking on a more regional character.

As an example of specific earthquake dangers to a municipality and the response of its government to them, we've chosen the state's largest and, after Sea-Tac, most densely populated area, the City of Seattle.

Adding to the danger from subduction zone quakes and from plate breakage deep under the Sound, the whole area, particularly Seattle, is threatened by a twenty-five-mile-long shallow fault. Appropriately named the Seattle fault, it could produce the region's worst possible damage because of its location. *(See Figure 6.1, next page.)*

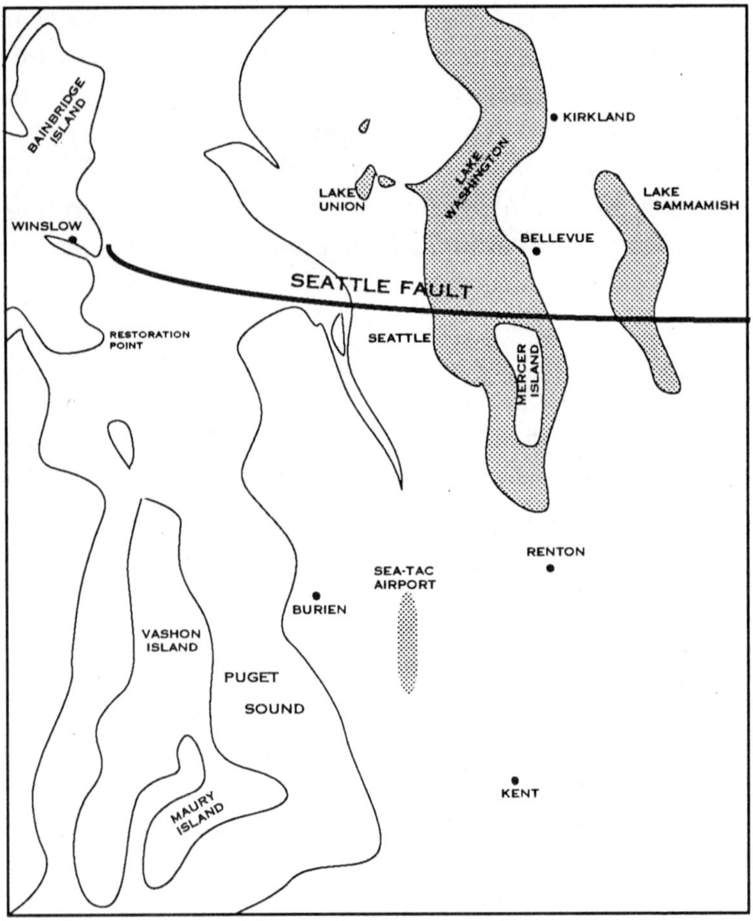

FIGURE 6.1. THE SEATTLE FAULT

Originating near Winslow, the fault heads east, passing under the full width of the Sound, under densely populated Seattle, then beneath Lake Washington and the northern tip of Mercer Island, and finally under and just past Lake Sammamish.

The frightening details of the history of the Seattle Fault have been unearthed only in the last few years. Estimated at magnitude 7 to 7.5, the last break, about 1,100 years ago, occurred along approximately ten miles of the fault, with some dramatic consequences. One side of the fault moved up over twenty feet, producing tsunamis of fifteen to twenty feet in the Sound and large destructive waves in the lakes. These waves, combined with the intense shaking, brought down large portions of the forest, particularly near the shoreline.

A similar quake in the future will be horrifying. Any structure on the fault section which breaks will be destroyed. Structures near the water will be inundated with powerful waves which will carry with them boats and whatever else is in the water at the time. Portions of the land will rise and others will drop, causing some areas to be permanently under water. Because faults near the surface cause lots of intense aftershocks, we can expect buildings weakened by the quake to be particularly susceptible to further shaking. Areas to the north and south will also receive heavy damage. For example, a Seattle fault quake would probably affect Vancouver, B.C., more than would a Cascadia subduction break.

Some Puget Sound readers may, at this point, be hoping that there is a large possibility of uncertainty in the data collected so far. Unfortunately, the geologic record is rather clear and someday the nightmarish events will almost certainly devastate Washington's most populous region. However, with adequate planning, and with some luck, the vast majority will survive and the area will be built anew—hopefully with safer structures.

One aspect of planning is engineering structures to deal with anticipated ground shaking. As was mentioned earlier, the general soil conditions in the Sound area are far from ideal. The specifics of potential soil movement in Seattle also doesn't look good. The 1965 quake was a real-life test of what will happen in the future.

With that tremor, there was significant landsliding.

Many hills in the city have an interesting and unfortunate layering of soils. The top portion of most hills rests on fairly impermeable clay deposits. As the area's abundant moisture filters through the porous upper soil, often sand, it reaches the densely packed clay and moves outward toward the hillside. The exposed soil becomes very wet and slides easily, even under non-quake conditions. The Queen Anne, Magnolia and Alki sections are all underlain with this soil configuration. The full length of the Seattle I-5 corridor is built along a known landslidable area and, during a very strong quake, may not only break up, but also be covered by landslides, significantly limiting transportation of relief materials.

Liquefaction was also a major problem in 1965. The area's glacial and river history has left deposits which are a mixture of gravel, silt and sand. All present and ancient river and stream valleys are soils of this general mix. Seattle also probably holds the title of "most moved around city" in the U.S. Like most cities with lots of shoreline, Seattle has extensive fill along rivers and bays. The downtown waterfront area and the Duwamish Delta industrial region are prime examples. But Harbor Island, the largest such fill in the world, and the Denny regrade, which both leveled and filled an enormous portion of the central city, significantly increased the land area which is very prone to liquefaction.

Add to this the potential weakening of the extensive abandoned coal mine tunnels under the city and the fact that Seattle sits on a bathtub-like basin which dramatically increases duration of shaking by trapping energy waves and bouncing them back and forth, and you have a very scary ground to be on during the Big One.

Because the problem has been known for some time, Seattle building code standards have been upgraded over the years, for example, in 1946, 1953 and 1964. In addition, new buildings downtown cannot now be above a certain height or larger than a certain square-footage and there is a ceiling on the number of building permits issued. Although there are many unreinforced brick and other dangerous buildings in the city, the code now requires retrofitting such structures when major upgrading and renovating is done.

The city has also initiated other measures to regulate development and increase safety in existing buildings. A booklet available from the Seattle Planning Department, "Seismic Hazards in Seattle," shows the areas of the city where soils are most likely to fail and the location of critical facilities in relation to them. Government officials can now more effectively administer life-saving land use planning. For example, developers in those areas prone to lots of liquefaction must first conduct soil studies to determine the best engineering.

A comprehensive proposal completed in 1994, The Seismic Hazard Reduction Plan, brings together all the city's programs to reduce earthquake danger. The plan develops priorities based on relative risk and vulnerability of structures. All city-owned bridges and buildings will be studied and necessary retrofitting will be completed by 1995, and water and light departments will upgrade their facilities.

Plans are also, of course, developed for response after a disaster. The city's Division of Emergency Management coordinates activities of various public agencies. To the extent that roads are passable, Seattle Metro will provide some of the necessary transportation and also communication, since many of the usual channels will be down.

Earthquake preparations are also, of course, being made in other cities throughout the Sound region. The second largest city has been actively involved for many years in a variety of activities. A Tacoma ordinance requires dangerous buildings to be demolished or repaired and retrofitted. Annually over one hundred buildings must meet this requirement. Tacoma's water reservoir system has also been completely upgraded. The city funds a position for an official who works on preparation and risk reduction with city workers and workplaces. This post also works with and educates the business community.

The city's schools have also been in the forefront of earthquake education, using attention-holding comic books and cartoon videos in the classroom.

The Sea-Tac Airport, although not a municipality, has its own police and fire departments. Because the airport is expected to take a big hit with a large quake and since getting the facility functional as soon as possible is so necessary for relief activities, the airport has developed and practiced an emergency management plan utilizing all the facility's employees.

Day by day, the Puget Sound region is getting more and better prepared for the disasters the future surely holds.

ROIL ON, COLUMBIA, ROIL ON *OR* IS DENIAL A RIVER IN WASHINGTON?

If a fault breaks underneath a body of water, whether it is an ocean, a lake, a river or artificially dammed, a large wave or series of waves may be created. If the quake faultline is under a dam or dike, the structure can collapse with the attendant release of flood waters.

The Columbia River is Washington's largest inland waterway. The amount of development along the Washington and Oregon banks has been truly dramatic over the past six decades. The river has been dammed to provide flood control, water for irrigation and electric power. And many industries which require lots of fresh water sit on its banks.

The collapse of a major dam can be one of the worst large-scale disasters imaginable. Usually there's not much time available for evacuation downriver after the break. Luckily, the chance of failure of a concrete dam like Bonneville is remote; of course, just because no concrete dam has ever failed as a result of earthquake shaking does not mean it cannot happen.

However, the Brownlee Dam on the Snake River near Baker City, Oregon, could conceivably fail as a result of a surface quake in that area. A major fault passes underneath the dam and it is not certain if the fault is active or not. Whether earthquake-related or not, the failure of this one dam could produce a wall of water that could destroy all the other downstream Snake dams. The Snake, of course, empties into the Columbia and although the flood waters would probably not do major damage to the Columbia's concrete dams, the water could spill over the top of these dams unless the water level was extremely low or was drawn down adequately before the floodwaters hit.

Major quakes in the Northwest could also touch off debris flows and giant landslides of unstable basalt and soil along the Columbia River Gorge and other rivers and lakes. Any resulting large waves could produce a lot of riverside and lakeside damage, and if roadways, bridges and communications were down, evacuation of communities would be very difficult.

The two largest Washington towns to be affected would be Vancouver in Clark County and Longview. At present, Clark County is part of the Portland Metro Preparation Area and hazard maps have been developed for the area.

Of even greater concern to some than the problems we've just mentioned is the matter of nuclear waste storage along the Columbia River. In eastern Washington, the Hanford Nuclear Reservation is considered the most dangerously polluted area in the country because of the large quantities of radioactive waste from plutonium production and other wastes which are stored there. Some of this hazardous waste has already leaked from storage facilities and there is fear that the effects of earthquake shaking could rupture other tanks, resulting in additional large-scale local and river radioactive pollution.

In addition, the only operating nuclear power plant in the Northwest, the Washington Public Power Supply System (WPPSS) Plant #2, located in the Hanford Reservation just north of Richland, raises the possibility of even more contamination. The fear is not just of distant subduction quakes, but also of local active faults. A 6.1 magnitude tremblor centered just south of Walla Walla shook up the area quite a bit in 1936. Two known active faults, the Saddle Mountain and Toppenish Ridge, are also near Hanford. Studies conducted for WPPSS claim that local quakes could not adversely affect the facility.

There is further concern about the reinforced concrete spent-fuel storage tank on the grounds of the shut-down Trojan nuclear power plant. Because there is no national repository for this highly radioactive waste, it was, through the life of the plant and well beyond the original intended capacity, stored on site. A cracked tank could prove disastrous. Studies commissioned by the owners have declared the tank earthquake-safe. Independent analysts concur, but some critics are not yet convinced.

Investigations in 1992 and 1993 of soils on the shores of islands in the Columbia uncovered conclusive evidence that ground shaking from the last subduction quake was strong enough to produce subsidence and liquefaction as far inland as an area ten miles east of Vancouver. One liquefaction site was near the Trojan spent fuel tank.

The debate over nuclear power, and the Trojan plant in particular, was very complex and emotional. Questions arose over the years about Trojan's design, construction and, most recently, age. One issue was the ground the facilities sat on. Although everyone agreed it is bedrock, there was disagreement as to its safety.

Studies conducted for PGE, the plant's majority owner, in 1988 and 1990 concluded that there were no potentially active faults under or very near the site and that local surface quakes or even the intensity of a very strong offshore subduction quake would not interfere with a safe shutdown of the plant. Others felt that the original studies of the site done before construction pointed to potentially dangerous faults which might cause future problems. Trojan never had on-site seismic monitoring equipment.

Over the years, the Nuclear Regulatory Commission (NRC) both supported and criticized Trojan's earthquake standards. In 1978 and 1979, the plant had to be shut down for a total of thirteen and a half months to correct some major design and construction deficiencies. Most recently, the NRC questioned the plant's safety systems and its ability to withstand a major quake.

Although the debate about the safety of the Trojan nuclear power generating facility is over, the issues raised about it and waste storage will be relevant to the future of the Pacific Northwest.

If there are radiation problems from any source after a quake, you should receive instructions on how to deal with them by listening to the emergency broadcast station on your portable radio.

ADDITIONAL RESOURCES

READING MATERIAL:

1) Yanev, Peter. Peace of Mind in Earthquake Country: How To Save Your Home and Life. San Francisco: Chronicle Books, 1991.
An excellent source of information for homeowners interested in making structural changes in their houses. Lots of photos and helpful diagrams. Also several chapters on other practical matters and on understanding earthquakes.

2) Nance, John J. On Shaky Ground: America's Earthquake Alert. New York: Avon Books, 1988.
An exciting, informative, well-written and hair-raising narrative of the 1964 Alaska earthquake, interspersed with lots of great, easy-to-understand general earthquake information. This best-seller is by a popular science writer who resides in Tacoma.

3) Walker, Bryce, and the Editors of Time-Life Books. Earthquake. Alexandria, VA: Time-Life Books, 1982.
Part of the "Planet Earth" series, this book has the best photos, drawings and diagrams you'll find anywhere. Superb text. Covers history, geology and the technology of prediction.

4) Harris, Stephen L. Agents of Chaos: Earthquakes, Volcanoes and Other Natural Disasters. Missoula, Montana: Mountain Press, 1990.
Another well-written book which explains California and Pacific Northwest geology in depth and our vulnerability to earthquakes and volcanoes.

5) Strengthening Wood Frame Houses for Earthquake Safety. Bay Area Regional Earthquake Preparedness Project (BAREPP), Oakland, CA (no date).
An excellent 32-page booklet to guide the homeowner through various do-it-yourself retrofitting projects. Each project includes a cost estimate. See BAREPP address and phone number in AGENCIES AND ORGANIZATIONS section below.

6) "Sunset's Guide To Help You Prepare For the Next Quake," Part 1. Sunset, October 1990, pages 163-177.
Very good article with great color photos on how to safeguard your home and possessions.

AGENCIES AND ORGANIZATIONS:

7) American Red Cross, Seattle–King County Chapter, 1900 25th Avenue S., Seattle, WA 98144-4708. (206) 323-2345.
King County residents can call or write to get information about disaster services. Also sells a popular first aid kit.

8) Mothers for H.E.L.P., T.L., P.O. Box 8399, Kirkland, WA 98034.
Citizen initiated, based and directed since 1988. "Help create prepared communities by increasing individual, family, neighborhood/community and school/business readiness for all types of disasters."

9) Washington Insurance Council, 13555 S.E. 36th Street, Suite 105, Bellevue, WA 98006-1485. (206) 747-7252.
Funded by the insurance industry in Washington state. Offers general information about earthquake insurance.

10) Bay Area Regional Earthquake Preparedness Project (BAREPP), Metrocenter, 101 8th Street, Suite 152, Oakland, CA 94607. (415) 893-0818.
Has a good selection of general and specialized publications. Write or call them and request their publications list.

11) Federal Emergency Management Agency (FEMA), Box 70274, Washington, D.C. 20024.
To receive a listing of FEMA publications related to disasters, write to the above address, ATTN: PUBLICATIONS, and request a copy of FEMA Publication Catalog (FEMA-20).

12) Department of Community Development, Emergency Management Division, Earthquake Preparedness Program, 4220 East Martin Way, P.O. Box 48346, Olympia, WA 98504-8346. (206) 459-9191.
Has a good selection of general and specialized Washington publications. Write or call them and request their publication list.

The best way to obtain information about what's happening in your community is to get in touch with your local Office of Emergency Management or Office of Emergency Services.

APPENDIX

INTRODUCTION

The remainder of this book is to be used after an earthquake. It is marked so you can turn to it immediately after a quake. This handbook is also bound so that you can turn to these relevant pages without any worry that the book will close on you. Some of the instructions refer to items in your earthquake preparedness kit.

When you actually do experience a quake, keep uppermost in your mind these important points:

1. Staying as calm as possible will significantly help your chances to successfully deal with the quake.
2. Remember the words most Californians know by heart: **Duck** (get under something safe), **cover** (your face and neck with your arms) and **hold on** (don't leave the safe spot until the shaking stops).
3. Do whatever seems right and safest and think through all your actions.
4. If you know what you're doing and others don't, shout instructions to them. Being an example can help prevent injury and death.

Also, every once in a while for the rest of your life—on the West Coast, at least—whether you are in a familiar or unfamiliar environment, stop and think and plan what you would do *right there* if an earthquake hit at that moment.

Of course, in the middle of a crisis, your thinking and actions can be very confused and not always as you had planned. But, in general, the more you know and the more you've practiced what you'd do during and after a quake, the greater your chances are of doing the right things.

So it is <u>very important</u> to read over <u>all</u> the information in these last sections <u>before a quake happens</u> until you are completely familiar and comfortable with everything.

Obviously, while a quake is happening, you won't be able to take out this handbook and read the WHAT TO DO DURING A QUAKE AND AFTERSHOCKS section. But you might review it quickly as soon as the primary shock is over because major aftershocks can hit in the minutes and days to follow. It is marked so you can turn to it immediately.

Now that you're working on your preparation, there's only one more thing to say:

"GOOD LUCK TO US ALL!"

EMERGENCY INFORMATION SECTION

A1

WHAT TO DO DURING A QUAKE AND AFTERSHOCKS

- KEEP CALM
- DON'T PANIC OR RUN
- IF OUTDOORS, STAY OUTDOORS
- IF INDOORS, STAY INDOORS

IN ANY HOUSE OR BUILDING

- **IF POSSIBLE**, WALK OR CRAWL CAREFULLY AND QUICKLY TO A CLOSE, SAFE PLACE
  - AWAY FROM:
    - ✖ WINDOWS
    - ✖ TALL FURNITURE
    - ✖ BOOKCASES
    - ✖ FILE CABINETS
    - ✖ ANY HEAVY OBJECTS THAT CAN FALL
    - ✖ ANY GLASS
    - ✖ OUTSIDE DOORS
    - ✖ BRICK WALLS AND FIREPLACES

A2

➡ **GET TO:**

- ✔ A HALLWAY AND LIE DOWN WITH YOUR FEET AND HANDS AGAINST THE WALLS

OR

➡ **GET UNDER:**

- ✔ A STURDY DESK OR TABLE AND HOLD ONTO THE LEGS

OR

➡ **GET NEXT TO:**

- ✔ AN INSIDE WALL AND COVER YOUR HEAD AND EYES WITH YOUR ARMS OR ANYTHING ELSE TO PROTECT YOUR HEAD

OR

➡ **STAND OR SIT:**

- ✔ IN AN INTERIOR DOORWAY (UNLESS THE DOOR IS HEAVY AND HITTING YOU OR CONTAINS GLASS)
- ✔ HOLD ONTO THE FRAME WITH BOTH HANDS

➡ **TURN YOUR HEAD AWAY FROM** ALL WINDOWS/GLASS/MIRRORS

☛ **EXIT CALMLY** WHEN THE SHAKING IS OVER

<u>DO NOT USE ELEVATORS</u>

IN YOUR HOME

☛ SEE PAGES A1 AND A2

PLUS:

☛ IN THE KITCHEN:
- ➨ **TURN OFF** THE STOVE
- ➨ **MOVE AWAY FROM** HEAVY APPLIANCES AND CUPBOARDS

☛ IN BED:
- ➨ **STAY** IN THE MIDDLE OF THE BED

✖ DO NOT HOLD PETS

✖ **DON'T RUSH TO ANYONE'S AID**—EVEN CHILDREN—UNTIL SHAKING STOPS. YELL THAT THEY SHOULD NOT MOVE AND THAT IT WILL BE OKAY SOON

✖ **DO NOT ATTEMPT** TO CATCH FALLING OBJECTS

IN FACTORIES & WAREHOUSES

☛ SEE PAGES A1 AND A2

PLUS:

☛ **MOVE AWAY FROM** HEAVY MACHINERY AND TALL SHELVES

IN A CROWDED STORE OR OFFICE BUILDING

- **SEE PAGES A1 AND A2**

PLUS:

- **MOVE AWAY FROM** GLASS DISPLAY CASES
- **IN A BRICK BUILDING**, GO TO A NON-BRICK INTERIOR WALL
- **STAY AWAY FROM** A SURGING CROWD
- **DON'T PANIC** IF FIRE ALARMS SOUND, LIGHTS GO OUT OR IF SPRINKLER SYSTEM IS ACTIVATED
- AFTER SHAKING STOPS:
  - ✔ USE EMERGENCY EXITS, BUT EXIT SLOWLY AND CALMLY
  - ✖ DON'T USE ELEVATORS
  - ➡ APPROACH STAIRS WITH CAUTION; THEY MAY BE DAMAGED

IN A THEATER OR STADIUM

- ☛ **STAY IN YOUR SEAT** AND PROTECT YOUR HEAD WITH YOUR ARMS OR A COAT
- ✖ **DON'T PANIC** IF FIRE ALARMS SOUND OR IF LIGHTS GO OUT
- ✖ **DON'T RUSH EXITS** WITH THE CROWD
- ☛ **WHEN SHAKING IS OVER,** USE REGULAR EXITS OR EMERGENCY EXITS. **DON'T USE ELEVATORS.** EXIT CALMLY AND SLOWLY.

IN A WHEELCHAIR

- ✖ **DON'T GET OUT**
- ➨ **MOVE** TO A SAFE, COVERED AREA
- ➨ **LOCK** THE WHEELS
- ➨ **COVER** YOUR HEAD WITH YOUR ARMS

IF OUTSIDE

➡ **MOVE AWAY FROM:**

- ✖ TALL TREES
- ✖ BRICK BUILDINGS
- ✖ ALL TALL BUILDINGS
- ✖ LARGE SIGNS
- ✖ STANDING AND DOWNED UTILITY POLES AND WIRES
- ✖ PLATE GLASS WINDOWS, GLASS DOORS

➡ **GET TO THE NEAREST:**

- ✔ NON-GLASS DOORWAYS OF SMALL, NON-BRICK BUILDINGS

OR

- ✔ LARGE, OPEN AREAS SUCH AS PARKS, ATHLETIC FIELDS, ETC.

✖ **DON'T RUN ALONG THE STREET**

IN A CAR

☛ **THE VEHICLE MAY BOUNCE A LOT**

☛ **PULL OVER** TO THE SIDE OF THE ROAD

☛ **HOWEVER,** IF YOU'RE:

➨ ON A BRIDGE

OR

➨ UNDER AN OVERPASS

OR

➨ VERY NEAR TALL BUILDINGS

OR

➨ UNDER WIRES

✔ **DRIVE** UNTIL YOU'RE IN A SAFE AREA AND STOP

✖ **DON'T GET OUT** OF THE CAR

☛ **KEEP YOUR SEATBELT** FASTENED

☛ **IF ANY WIRES FALL ON THE VEHICLE,** ROLL UP THE WINDOWS

☛ **WHEN YOU DO DRIVE AWAY:**

✖ **DON'T DRIVE ON OR UNDER** ANY DAMAGED ROADWAY WHICH MIGHT COLLAPSE

➨ **WATCH OUT FOR DOWNED WIRES**

WHAT TO DO AFTER A QUAKE

(IF YOU'RE ON THE COAST, FIRST SEE PAGE B16)

DO THESE THINGS FIRST:

- ➡ CHECK YOURSELF AND OTHERS FOR INJURIES. IF NECESSARY, ADMINISTER FIRST AID. USE YOUR FIRST AID BOOK AND, IF POSSIBLE, WEAR LATEX GLOVES. FOR RESCUE, WEAR HEAVY WORK GLOVES AND THICK-SOLED SHOES OR BOOTS AND USE A CROWBAR TO HELP LIFT HEAVY OBJECTS OFF PEOPLE.

IF THE STRUCTURE YOU ARE IN IS SO WEAKENED THAT IT MIGHT COLLAPSE, EXIT CALMLY AND CAREFULLY. AFTERSHOCKS MIGHT BRING THE STRUCTURE DOWN.

IF THE STRUCTURE APPEARS SOUND, YOU SHOULD NEXT TRY TO **PREVENT AND CONTROL FIRES.** BUT REALIZE THAT AFTERSHOCKS WILL PROBABLY OCCUR, SO THINK ABOUT WHAT WOULD BE A SAFER AREA TO GO TO WHEN THEY HIT. DO NOT RISK LIFE TO PROTECT PROPERTY.

- **TO STOP A FIRE:** IF A ***SMALL*** FIRE IS IN PROGRESS AND YOU BELIEVE IT IS CONTAINABLE:
  - **PUT ON** THICK-SOLED SHOES OR BOOTS.
  - **EVACUATE** CHILDREN, THE AGED AND THE HANDICAPPED FROM THE STRUCTURE.
  - **YELL FOR HELP** AND PULL ANY FIRE ALARMS.
  - **IF POSSIBLE,** TURN OFF GAS AND ELECTRICITY (SEE PAGES B3 THROUGH B6).
  - **USE SAND** OR A CHEMICAL FIRE EXTINGUISHER OR WATER HOSES TO PUT OUT THE FIRE.
    - **FOLLOW CAREFULLY** ANY WRITTEN DIRECTIONS ON THE EXTINGUISHER.
    - **ALWAYS SPRAY BACK AND FORTH** AT THE BASE OF THE FIRE.
  - ✖ **DON'T USE WATER FOR AN ELECTRICAL FIRE** UNLESS ELECTRICITY HAS BEEN TURNED OFF.

- **GREASE FIRES** ON STOVES CAN BE EXTINGUISHED WITH SALT OR BAKING SODA OR BY SMOTHERING WITH A METAL LID OR WITH A FIRE EXTINGUISHER, BUT NOT WITH WATER.

- **IF YOU CAN'T CONTAIN THE FIRE,** LEAVE THE STRUCTURE AND CALL THE FIRE DEPARTMENT, IF POSSIBLE.

- **REMEMBER, FIRES MOVE RAPIDLY** AND CAN GET OUT OF CONTROL QUICKLY. **DON'T TAKE CHANCES!**

☛ **TO PREVENT FIRES:**

- **TURN OFF STOVE AND OVEN.**

- **TURN OFF THE MAIN GAS VALVE** (NEAR THE GAS METER) WITH A CRESCENT WRENCH IF IT SEEMS THERE IS A GAS LEAK ANYWHERE AND **YOU CAN SMELL GAS OR IF YOU HEAR A HISSING SOUND OF GAS ESCAPING.** *(See Figure B.1, next page.)*

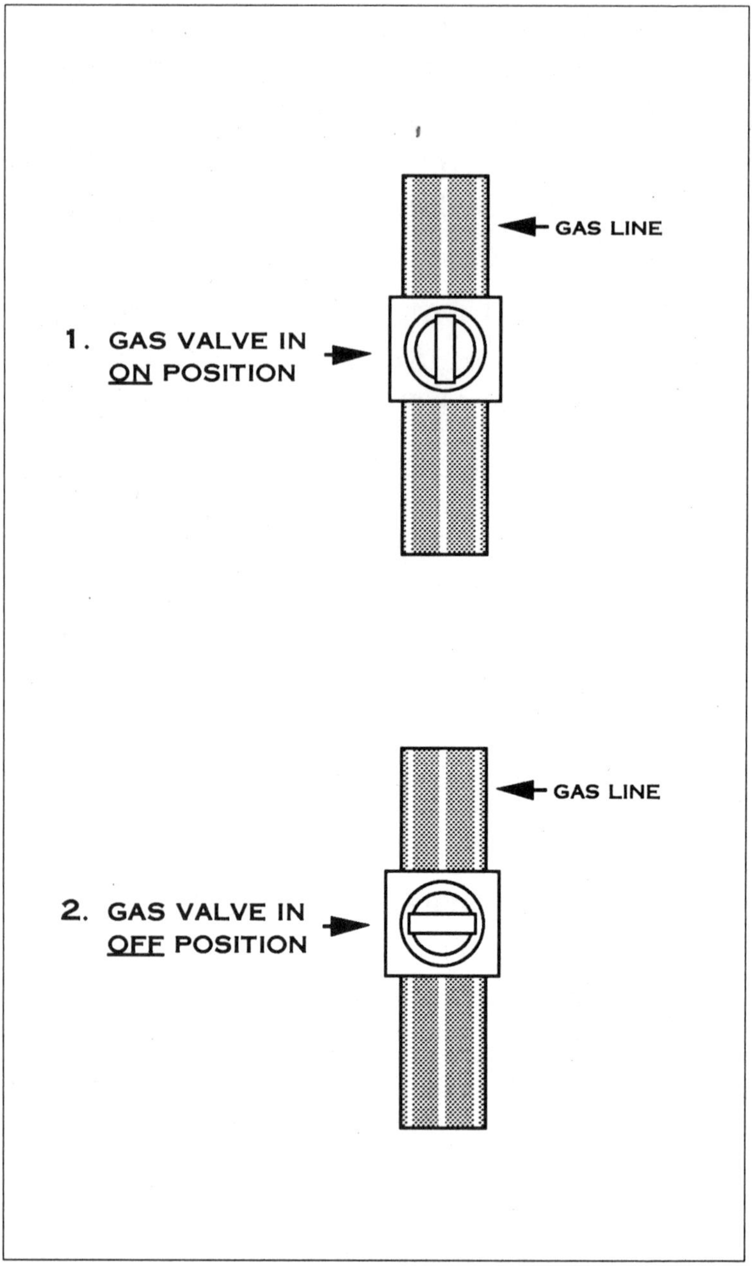

FIGURE B.1. HOW TO TURN OFF MAIN GAS VALVE

- ✖ **DON'T USE ANYTHING ELECTRICAL** OR MATCHES OR OTHER FLAMES IF ANY POSSIBILITY OF GAS LEAKAGE EXISTS.

- ➡ **IF GAS IS LEAKING INSIDE** OR MIGHT LEAK FROM LINES OUTSIDE THE STRUCTURE, TURN OFF THE ELECTRICITY AT THE MAIN BOX *(see Figure B.2, next page)*. IF YOU CAN, OPEN WINDOWS, THEN LEAVE THE HOUSE. NOTIFY NEIGHBORS AND PLACE WARNING SIGNS NEAR THE STRUCTURE. IF POSSIBLE, CALL THE FIRE DEPARTMENT. DON'T GO BACK INTO THE STRUCTURE UNTIL THE LEAK HAS BEEN TAKEN CARE OF.

- ➡ IF THERE APPEARS TO BE NO GAS LEAKAGE, NEXT CHECK THE ELECTRICAL WIRING. IF YOU SEE OR HEAR LIVE OR SHORTING WIRES, TURN OFF THE POWER AT THE MAIN BOX. *(See Figure B.2, next page.)*

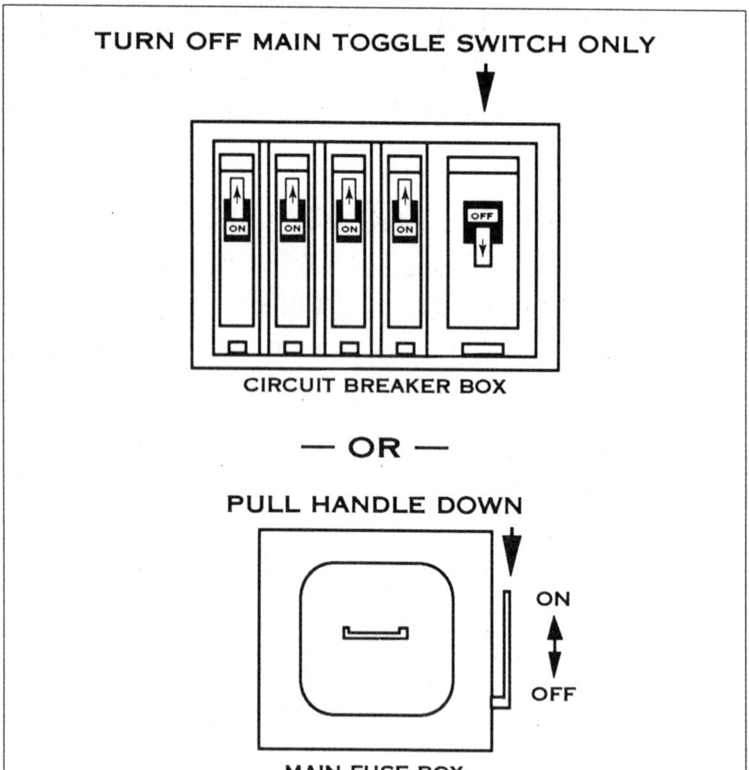

FIGURE B.2. HOW TO TURN OFF THE ELECTRICAL POWER AT MAIN BOX

DO NOT USE YOUR PHONE OR ANY PHONE UNLESS YOU HAVE AN EMERGENCY. IT TIES UP THE LINES AND BLOCKS CALLS OF PEOPLE WHO DO HAVE EMERGENCIES.

- **IF YOU HAVE AN EMERGENCY,** BUT DO NOT HAVE USE OF A PHONE, GET TO A HOSPITAL, POLICE STATION OR FIRE STATION, WHICHEVER IS CLOSEST. GET THERE FAST, BUT SAFELY. IF YOU CAN'T DO THIS, ASK HELP FROM YOUR NEIGHBORS AND/OR ATTRACT THE ATTENTION OF ANY DISASTER RELIEF PLANES OR VEHICLES YOU MIGHT SEE.

IF YOUR WATER PIPES ARE DAMAGED AND LEAKING, TURN OFF THE WATER AT THE MAIN VALVE.

AFTER YOU HAVE SECURED YOUR HOUSE OR BUILDING OR HAVE ABANDONED IT, SEE IF THERE'S ANYONE IN YOUR IMMEDIATE AREA WHO NEEDS HELP.

With the worst dangers taken care of, you must now think about food, water, shelter and sanitation. However, there are some other very important things to do and to be aware of:

- ➡ **LISTEN TO YOUR PORTABLE RADIO** FOR INFORMATION AND DIRECTIVES. FOLLOW ANY DIRECTIONS AND BE PREPARED FOR EVACUATION AT ANY TIME.

- ➡ **IF YOU'RE TOLD TO GO TO A NEARBY EVACUATION AREA, DO THIS.** DO NOT TRY TO DRIVE FAR AWAY IN THE HOPES THAT YOU'LL FIND A SAFER, UNDAMAGED AREA. YOU MAY GET STUCK IN A TRAFFIC JAM, COME INTO CONTACT WITH LIVE WIRES OR UNSAFE ROAD SURFACES OR BE HIT BY A PANICKED DRIVER.

- ➡ **DO NOT LEAVE CHILDREN.** HOLD THEM AND REASSURE THEM.

- ✖ **DO NOT, FOR ANY REASON, COME INTO CONTACT WITH DOWNED WIRES** AND UTILITY POLES OR ANYTHING WHICH IS TOUCHING THEM.

- ➡ **STAY AWAY FROM:**
  - ✖ ANY AREA WHICH MIGHT LANDSLIDE
  - ✖ DAMAGED BUILDINGS
  - ✖ ANYTHING ELSE WHICH MIGHT FALL ON YOU

- ✖ **DON'T WANDER AROUND TO VIEW DAMAGE.** Your presence on streets may only slow down or endanger relief efforts. In addition, you only increase the likelihood that you'll come into contact with live wires or enraged animals.

➡ **PETS:** Keep dogs on leashes and indoors, if possible. Cats may hide for a while, but will probably return home. **KEEP YOUR DISTANCE FROM PACKS OF DOGS.**

This is a good time to look over the previous section, WHAT TO DO DURING A QUAKE AND AFTERSHOCKS, so you'll be able to deal with aftershocks.

SHELTER

✖ **IF THE STRUCTURE HAS BEEN WEAKENED** AND MIGHT COLLAPSE WITH AFTERSHOCKS, DON'T GO BACK IN UNLESS IT'S ABSOLUTELY ESSENTIAL.

➡ ANY SHELTER SHOULD BE **AWAY FROM TREES, POWER LINES AND BUILDINGS** WHICH COULD FALL ON IT. IDEAL AREAS ARE OPEN FIELDS AND PARKS, SCHOOL YARDS AND ATHLETIC FIELDS.

☛ **LIVE IN:**

✔ **A CAMPER OR AN RV OR A CAR.** They're designed to move with shocks and they're not likely to tip over. Whenever possible, use seatbelts.

OR

✔ Live in a **TENT.**

WATER

- ☛ **IT IS VERY IMPORTANT TO NOT BECOME DEHYDRATED.** IT'S ALSO VERY IMPORTANT THAT THE DRINKING WATER AND OTHER LIQUIDS BE SAFE, SO YOU DON'T BECOME SICK WITH DYSENTERY OR VOMITING WHICH CAN DEHYDRATE YOU.

- ☛ **IF THERE IS DAMAGE TO SEWER AND WATER LINES,** IT IS POSSIBLE THAT SEWAGE COULD CONTAMINATE YOUR TAP WATER.

- ☛ **SAFE DRINKING WATER** AND LIQUIDS CAN BE OBTAINED FROM:

  - ✔ Stored bottled water
  - ✔ Ice cubes
  - ✔ Canned foods liquids (glass containers may be damaged—filter with handkerchief or other cloth before drinking)
  - ✔ Toilet tanks (NOT BOWLS AND NOT IF WATER IS CHEMICALLY TREATED TO KEEP BOWL CLEAN).
    - ✖ **DON'T FLUSH TOILETS** UNTIL YOU ARE SURE SEWER AND WATER LINES ARE INTACT.
  - ✔ Hot water heaters. FILTER WATER WITH HANDKERCHIEF OR OTHER CLOTH BECAUSE OF THE POSSIBILITY OF GLASS.

- To drain water from hot water heater, follow these directions:

  1. Make sure gas or electricity (whichever heats the water) is turned off.
  2. Turn off the valve which lets water into the tank.
  3. Turn on nearby hot water faucets.
  4. Open the valve at the very bottom of the tank. Drain water into a clean container, using a hose, if necessary.
  5. Discard any rusty water.

TO PURIFY ANY WATER FOR DRINKING:

- Use IODINE TABLETS OR CHLORINE TABLETS. Follow directions on the bottle

OR

- Using an eyedropper, add 2 drops of chlorine bleach (or 3 drops of iodine) to each quart of water (4 drops of chlorine or 6 drops of iodine if water is cloudy), mix and let stand for 30 minutes. If, after 30 minutes, there is no chlorine smell to the water, it is still not safe, so repeat the process.

FOOD

➡ **EAT ALL FOODS WHICH WILL SPOIL** SOON FIRST, NEXT REFRIGERATED FOODS, THEN FROZEN FOODS AND LASTLY, DRIED AND CANNED FOODS.

☛ If you keep refrigerators and freezers closed, food will keep for several days after the electricity is off. Food which does not smell bad and feels cool is probably good. Perishable food is not safe if it has been at room temperature after a couple of hours.

☛ It's best to cook food outside on a camper stove, hibachi, etc. Use an indoor fireplace only if you are <u>absolutely certain</u> there are no gas leaks and that your chimney is undamaged. Examine the chimney outside for damage, keeping enough distance so it can't fall on you. However, keep in mind that you may not be able to detect interior damage to your chimney.

SANITATION

➨ **IF YOU SEE OR SMELL BROKEN SEWER LINES** ANYWHERE IN YOUR AREA, DO NOT FLUSH YOUR TOILET—YOU WILL ONLY DEPOSIT YOUR WASTE NEARBY WHERE IT CAN BE VERY DANGEROUS.

☛ **IF YOU CAN'T USE YOUR TOILET:**

- ➨ Dig a hole or trench in your yard several feet deep as a toilet. Every time it is used, throw in some lime or disinfectant. Keep the hole covered when not in use. When full, cover with dirt, then with wood.

OR

- ➨ Line your toilet bowl with a plastic bag. After using it, seal it and bury it.

➨ Bury all garbage.

➨ Plug sink and bathtub drains to stop backup from sewers.

MISSING FAMILY MEMBERS AND FRIENDS

Get in touch with the Red Cross for information on the health and welfare of family members only. But note: it will take about 3 days for the Red Cross to activate this service. If you must evacuate your home, leave a note on your front door telling friends of your whereabouts.

AT THIS MOMENT, THINGS MIGHT SEEM VERY BAD, BUT REALIZE THAT DISASTER RELIEF EFFORTS ARE UNDERWAY LOCALLY, REGIONALLY, NATIONALLY AND INTERNATIONALLY. YOU WILL MAKE IT OUT OF THIS.

DOING SOMETHING CONSTRUCTIVE CAN HELP YOUR SPIRITS NOW. If you have things that can be cleaned up, do this. Everyone should wear gloves and thick-soled shoes and boots. Any dangerous chemical materials that have spilled can be buried or put in tight containers.

Debris like glass can be moved so it is less of a danger.

When opening cupboards or doors of any storage area, do so very carefully.

Board up broken windows and post signs warning of any dangers.

Make sure all receivers are placed back on phones. This helps the system return to normal and allows lines to be used for emergency calls.

Involve older children in safe clean-up activities. This will allow them to also feel they have some control and will help reduce some of their fears. Very young children should be held and comforted. Children should not be left alone.

For insurance and tax purposes, make lists now (or soon) of all possessions which were damaged or destroyed. If possible, photograph the destruction before clean-up.

Be a good Samaritan: Help out others in your neighborhood, particularly the aged, handicapped, etc. Make sure others learn the survival information listed in this Appendix.

AFTER SEVERAL DAYS YOU SHOULD, IF POSSIBLE, MAKE EVERY ATTEMPT FOR A SEMBLANCE OF NORMAL LIFE. When your home has been declared safe to live in and the danger is officially over, clean up your house and yard and sleep inside again. Talk about what has happened and your fears with your children and neighbors and encourage them to do the same.

WHAT TO DO AFTER A QUAKE ON THE COAST AND COASTAL BAYS

- ☛ ONE OF YOUR BIGGEST DANGERS NOW IS THE POSSIBILITY OF A TSUNAMI (TIDAL WAVE).
- ☛ IF YOU'RE ON LOW-LYING LAND AND YOU FEEL AN EARTHQUAKE, **GET TO HIGH, SOLID GROUND QUICKLY AND SAFELY** AS SOON AS POSSIBLE. IF YOU CAN'T DRIVE, RUN!
- ☛ **IF YOU HAVE PREPARED AN EMERGENCY EARTHQUAKE BACKPACK,** TAKE IT WITH YOU.
- ☛ **DO NOT WASTE TIME** BY TRYING TO SAVE VALUABLES OR SENTIMENTAL POSSESSIONS.
- ☛ A TSUNAMI IS USUALLY MORE THAN ONE BIG WAVE. **DON'T ASSUME THAT AFTER A BIG WAVE HITS, THE DANGER IS OVER.**
- ☛ **TURN ON YOUR PORTABLE OR CAR RADIO** AND FOLLOW ALL EMERGENCY INSTRUCTIONS. DO NOT RETURN TO YOUR HOME, WORKPLACE, HOTEL OR LOW-LYING AREAS UNTIL AUTHORITIES HAVE ISSUED AN ALL-CLEAR.
- ☛ GO IMMEDIATELY TO HIGH GROUND NOW!

ADDITIONAL COPIES of The Washington Earthquake Handbook or the The Oregon Earthquake Handbook may be obtained by sending a check or money order along with the order form to the address below.

Each copy is $11.95, postpaid. A generous discount is available for quantities of 20 copies or more. Write to the address below for particulars.

ORDER FORM

- ❑ Enclosed is $_______ for ____copies of The Washington Earthquake Handbook @ $11.95 each
- ❑ Enclosed is $_______ for ____copies of The Oregon Earthquake Handbook @ $11.95 each

Total amount enclosed: ________

NAME ____

ADDRESS ______________________________________

CITY __________________ STATE ___ ZIP ________

Make checks payable to: "Vern Cope/Handbook"
Allow 2 to 3 weeks for delivery
Mail order form and check or money order to:

Vern Cope
Earthquake Handbook
P.O. Box 19843
Portland, OR 97280